HOW TO CONDUCT YOUR OWN SURVEY

HOW TO CONDUCT YOUR OWN SURVEY

PRISCILLA SALANT
and
DON A. DILLMAN

John Wiley & Sons, Inc.
New York • Chichester • Brisbane • Toronto • Singapore

This text is printed on acid-free paper.

Published by John Wiley & Sons, Inc.

Library of Congress Cataloging-in-Publication Data:

Salant, Priscilla.
How to conduct your own survey/Priscilla Salant and Don A. Dillman.
p. cm.
Includes bibliographical references and index.
ISBN 0-471-01267-X (cloth)
ISBN 0-471-01273-4 (paper)
1. Social sciences–Research. 2. Social surveys. I. Dillman, Don A.
II. Title.
H62.S3185 1994
300'.723–dc20 94-5636

Printed in the United States of America.

10 9 8 7 6 5

This book is dedicated to

Richard S. Salant
for the standards he set in
honest, balanced reporting

and

J. Orville Young
for his gentle but persistent reminders that
the ultimate goal of research and the land grant university
is to help people.

Acknowledgments

We wish to acknowledge, with thanks, the support of several organizations and many individuals who made this book possible.

Major grant support was provided by the Rural Poverty and Resources Program of the Ford Foundation with assistance from the Rural Economic Policy Program of the Aspen Institute. The vision of people from both Ford and Aspen led to this effort in the first place, and to its completion. Additional support was provided by the Social and Economic Sciences Research Center, the Department of Rural Sociology, and the Department of Agricultural Economics at Washington State University.

Very conscientious reviewers gave us critical help and detailed suggestions along the way. They include Helen Berg, Ellen Golden, Stephen Heeringa, Howard Ladewig, Carol Lamm, Robert Mason, and Todd Rockwood. Any remaining errors, omissions, or lack of clarity are our own responsibility. We also owe a great debt to people who contributed in many ways, large and small: Rita Koontz, Deborah Schwenson, Tammy Small, John Tarnai, Jonathon Vlaming, and Anita Waller.

Our biggest debt of all is to the hundreds of people and organizations who provided the experiences and inspiration to tackle this book. We gained valuable practice as we responded to their requests for help. In addition, they convinced us of the need for a book that describes surveys in plain English. Many of these people and organizations appear in examples cited in this book, although details have been changed to preserve anonymity.

Contents

Preface

This book is a response to hundreds of calls for help that we have received over a period of many years in our work as survey consultants. The calls have come from diverse sources—county commissioners, businesspeople, graduate students, artists, officers of voluntary associations, university professors, and concerned citizens.

People have asked for help in designing and implementing surveys, as well as in analyzing and interpreting survey results. For example, we've been asked:

- Is a 3 percent sample of local residents enough for a valid survey?
- Can we use the telephone directory to draw our sample?
- Is a 55 percent response rate enough for accurate results on a mail survey?
- Can we trust results from a telephone interview more than a mail questionnaire?
- Isn't an agree-disagree format better than an open-ended question to which people have to think up their own answer?
- A survey research consultant said our project would cost $50,000. *Isn't there some alternative?!*

Our answers to questions like these often start out, "Well, that depends. . . . " Despite our best intentions to be brief and unambiguous, our answers are often long and complex. Sometimes, it's because we use technical words—terms like *ordinal scaling, variance,* and *confidence limits.* Other times we raise issues our caller hasn't thought about. And as we continue to introduce new language and ideas, we all get a little frustrated. We are reminded of the person who went into a hardware store to buy a rake and found herself listening to an unwanted and complicated discussion of various mulching techniques. In the end, neither the customer nor the would-be teacher of mulching methods was satisfied.

We, too, are frustrated during encounters like these, for several reasons. On one hand, surveys can be useful and powerful tools for both public and private organizations that need to know the characteristics and opinions of people they serve. Surveys are often the only means of finding out the percentage of people who own their homes, who commute more than two miles to work, or who have some other characteristic of interest. And to get this information, it is often necessary to survey only a *small* fraction of the population. Therein lies the potential power of sample surveys.

On the other hand, how and why surveys provide accurate information is far from obvious. For example:

- Getting 200 responses in a survey of 250 randomly sampled people may produce far more accurate information about the general public than 2000 responses from a sample of 10,000.
- The number of respondents needed to achieve a certain level of precision in a city or county survey is often about the same as the number needed in a national survey.
- Sometimes the benefits of sampling are minimal, and it makes more sense to survey everyone in a particular population.
- Even though a survey gets a 95 percent response rate, it may produce inaccurate and worthless results, the victim of other types of errors.
- Changing the order of certain questions can dramatically affect—perhaps by 20 to 30 percent—the number of people choosing certain answer categories.

In times past, this book would have been considered unwise. Surveys were regarded as too complex and costly to be conducted by nonprofessional surveyors. In the last two decades, however, doing surveys by mail, telephone, and the drop-off method has come of age. They are now viable alternatives to the once standard and expensive face-to-face interview. These developments have made the task of designing and implementing surveys harder in some ways and simpler in others—harder because people now have more methods from which to choose and simpler because alternative methods can sometimes be used with relative ease. It is now possible for people with little research experience to do credible surveys.

Our first goal here is to bring the process of doing surveys within reach of people who need survey information but have no formal survey training. Rather than writing for professional researchers, who deal with complex national surveys and computer-assisted interviewing, we have developed this book for a different kind of reader. Readers who would find this book useful could include the citizen who is measuring support for a community crime-prevention program, the medical doctor who is surveying patients, and the town mayor who is gauging public opinion on a local bond issue.

Our intended audience also includes graduate students and professors in both the natural and social sciences. Students often do relatively straightforward surveys as part of their master's and doctoral work, and many professors do the same in the course of carrying out their research responsibilities. Our experience suggests that both could benefit from an applied, nonacademic guide to doing surveys. Hence, this book explains how surveys work and what makes them accurate. Step by step, we describe decisions that must be made in the course of a survey, including whether a survey is even appropriate, and if so,

- whether mail, telephone, or face-to-face interviews will work best;
- whether and how to draw a sample;
- how to write a questionnaire;
- how to implement the survey; and
- how to compile and report results.

We've tried to translate technical, scientific survey concepts into ordinary language, avoiding wherever possible jargon used by statisticians, sociologists, and psychologists. We do not reject the concepts of these disciplines as irrelevant. Indeed, we have relied on them in developing our book, and we view them as critical to the science that underlies what we have written. However, we believe that useful, small-scale surveys can often be done without invoking the complexity inherent in national, general-public surveys, where great precision is essential. Thus, we emphasize the simple designs and straightforward solutions that our particular audience can use.

The rapidly changing society in which we live produces a demand for survey information that cannot easily be ignored. For example, proposals to modify how a city agency does business, offers a new service, or obtains outside funding often require concrete evidence that a problem exists. In this last decade of the twentieth century, it is no longer adequate to justify a new city-wide crime program with statistics for the nation as a whole or even the state. Local data are a prerequisite. It is this need that has prompted so many people from small organizations and jurisdictions to call us for help.

To meet our primary goal of writing for nonprofessional surveyors, we discuss what surveys can and cannot do, how various sources of error can undermine accuracy, and ways to solve major problems of survey design. However, we fully expect that, for some readers, the procedures we propose will be insufficient. The kind of information they need or the kind of population they survey will require more complex survey designs. Hence, our second goal is to help people recognize when they need more help and to point them in the right direction. Understanding when certain aspects of a survey are too complicated for the reader to attempt without help is as important as understanding what one *can* do without help. Timely consultation may mean the difference between success and failure.

It is likely that some readers will supplement this book with other, more detailed texts. (We suggest references at the end of each chapter.) Others may decide to get professional assistance in conducting their surveys. In fact, we recommend one or both options when circumstances call for anything but a very simple survey design. Regardless of whether you decide to go it alone or hire a consultant, you can increase your chances of success by using this book to become more informed about what makes a good survey.

Among the people who inspired us to write this book was a city official who was trying to decide whether to spend $25,000 of public money on new downtown street signs. He lamented trying to commission a survey, only to be told it would cost $50,000, twice the amount he proposed spending on the signs. We hope that those who read this book will gain insight into survey methods proportionate to their information needs—methods that they can implement themselves. In addition, we hope readers will learn to recognize when their information needs pose more complex problems and when, in certain instances, the only satisfactory solution may be not to do the survey at all.

1

Practical Surveys

The woman on the telephone sounded anxious. As the mayor of a nearby town, she had appointed a task force to resolve conflicts over how to use a vacant lot in the central business district. The task force had recommended doing a survey to decide whether the space should be used for more parking or for a downtown plaza. The mayor hadn't budgeted money for such a survey but wanted to support the task force. We scheduled a meeting for the next morning.

The mayor arrived at our office in much better spirits. She announced that the problem was solved:

> We have some high school students who work for us 1:00–4:00 P.M., Tuesdays and Thursdays. We'll give each of them a clipboard with one question written at the top: "In your opinion, what should be done with the vacant lot next to the Perkins Building?" We'll have these interns stand at downtown street corners, stop everyone who comes by, and tally their answers until we get about two hundred. That should be plenty; after all, surveys covering the whole country often interview only a thousand.

A professional survey methodologist would have been justified in pointing out serious problems with the mayor's proposal. She ignored many of the fundamentals of good surveying. The sample would have been biased because it included only people who happened to show up downtown in the early afternoon. The question was too vague to get clear answers, and the mayor's reasoning on sample size was incorrect. No one would have been able to interpret the results for any purpose whatsoever.

Although we could have responded that way to the mayor, we chose not to. Instead, we talked with her about how and why surveys work, the first topic of this chapter. Next, we briefly discussed the issues covered in the rest of this book, and in particular, what causes survey errors. Finally, we offered some specifics about how to design a survey to meet the goals of both the mayor and the task force: to spend a reasonable amount of money finding out what the majority of citizens wanted to do with the vacant lot.

Our solution involved conducting a mail survey of randomly selected registered voters. The questionnaire consisted of an attractive, four-page booklet and included several questions that would enable the task force to interpret opinions of different interest groups. The mayor's student interns provided most of the labor, still on Tuesday and Thursday afternoons, with additional help from the task force. Envelopes and letterhead were donated, and a statistics instructor from the local community college did the analysis. For about $1200 (mostly for postage), the data were collected, the results analyzed, and a report made to the task force. Under the mayor's direction, a very practical and useful survey was conducted at low cost. Our purpose in this book is to describe when, why, and how to conduct similar surveys.

If you have a practical survey problem, you aren't alone. Here are three more examples:

> An economic development council in the rural Northwest wanted to know people's preferences for various development goals. The council used a mail questionnaire to ask local residents to compare the importance of jobs and environmental quality. Other questions asked for opinions on such development strategies as tourism growth, prison construction, and hydroelectric projects.
>
> An art museum in a large northeastern city conducted face-to-face surveys with people attending an exhibit. Respondents were randomly selected as they left the exhibit and were asked how often they attended the museum, what types of exhibits they preferred, and whether they would pay an annual membership fee. The basic purpose of the survey was to learn what patrons wanted to see and how much they would pay for it.
>
> A state employment agency in the upper Midwest used a telephone survey to evaluate the effectiveness of its worker retraining program. People who participated in the program were asked whether retraining had helped them find a job, and, if so, how long had they been working, how steady had their job been, and what wages had they received.

The people who did these surveys wanted to find out the characteristics, behaviors, or opinions of a particular population. They hoped to gather this information by interviewing only part of the total group in which they were interested. Hence, we say they conducted **sample surveys.**

Doing accurate surveys means answering a series of key questions, such as:

- How many people are required for a valid survey?

- Should the survey be conducted through the mail, by telephone, or face-to-face?
- Should everyone have a chance to answer the questionnaire or only a select group?
- How should the sample be selected?
- How high should the response rate be?
- How accurate will the results be?

Our goal is to help you answer these kinds of questions and enable you to take advantage of opportunities offered by surveys.

Why Surveys Work

The three surveys described above were guided by the same basic principles as national-level surveys carried out by well-known polling organizations. Consider, for example, the presidential preference polls conducted in the final days of the 1992 campaign. Figure 1.1 shows the results of seven polls taken at various times between October 28 and November 2. Each polling organization applied its own definition of likely, or probable, voters, drew its own sample, and conducted its own telephone interviews. Even so, the results were quite similar. Clinton's support was estimated at 43 to 45 percent, Bush's at 35 to 39 percent, and Perot's at 14 to 18 percent. When the final vote was tallied, Clinton received 43 percent, Bush 38, and Perot 19. Only the estimates of Perot's supporters were off by more than 2 or

Figure 1.1
Preference polls came close to estimating the final vote in the 1992 presidential election.

	Clinton %	Bush %	Perot %	Other/ Undecided %	Sample size
Actual vote on Nov. 3	43	38	19	—	—
Preelection polls					
Gallup	44	37	14	5	1589
CBS	45	37	15	3	1731
NBC/*Wall Street Journal*	44	36	15	5	806
Harris	43	37	16	4	1975
Tarrance	43	39	18	—	789
New York Times/CBS	44	35	15	6	2086
Washington Post	43	35	16	6	722

Polls were conducted at various times during the period Oct. 28–Nov. 2, 1992.

Source: The Polling Report, Inc.

3 percentage points, probably reflecting the difficulties of measuring undecided voters in a three-way race.

How could seven pollsters, acting independently and interviewing such small samples of likely voters, project the support for Clinton and Bush so closely? To answer this question it is necessary to understand the principles of statistical sampling.

The purpose of a sample survey is to obtain information from a few respondents in order to describe the characteristics of hundreds, thousands, or even millions. The sample is selected in such a way that it represents the entire population; that is, observations from the sample can be generalized to the larger group. However, estimates based on samples are subject to **sampling error,** a potential error that occurs when surveyors interview only a sample, instead of the entire population.

In the polls cited in Figure 1.1, the samples consisted of between 722 and 2086 adults, while the population consisted of about one hundred million likely voters. The sampling error could have been as large as 3 or 4 percent in either direction. (The actual magnitude depended on sample size.)

Polling organizations can stay in business only if they have a good reputation for making accurate projections at the lowest possible cost. To do so, they use complex and sophisticated techniques to ensure that their samples are representative of the population and, therefore, that their projections are as accurate as possible. Although these techniques are far more complicated than any that local researchers might need to use in their own community or state, the principle is the same: *Surveys can be used in a scientific way to realize the great benefits of interviewing a representative sample instead of the whole population.*

Polling organizations save time and money by interviewing only one or two thousand people during the week before a presidential election instead of all one hundred million who are likely to vote. Let's look closely at why pollsters can make election predictions by surveying such a small proportion of the population.

Part of the explanation has to do with how we view people. Skeptics of surveys and polls often observe that every individual is unique, and therefore we cannot possibly draw a sample of a few to represent the many. One has only to look around a crowded airport or auditorium and to try finding just two identical people and the assertion seems well supported. People *are* unique!

However, when we are interested in doing a survey, we begin by looking at people somewhat differently. Instead of focusing on the totality of each individual, we change our focus to look at only the

How Do Major Polling Organizations Select Their Samples?

As an example, consider the procedure used by the Gallup organization. Gallup researchers begin by dividing the entire United States into very small segments of land area and then stratifying, or classifying, these segments according to geographic area and size of community. Next, in two separate stages, they select a sample of 350 segments in such a way that the sample reflects the U.S. as a whole, again in terms of geographic area and size of community. And finally, within each area, they randomly select telephone numbers for their interviews.

Gallup conducts many other surveys in addition to its election polls. Topics on which the company surveyed Americans in early 1992 included people's perception of the U.S. tax system, their response to negative campaign ads, and their plans to watch the Super Bowl.

characteristics in which we are particularly interested. For example, if we want to know what proportion of voters in a community favors using tax breaks to attract new manufacturing businesses, we shift our interest to one aspect of people's lives—Do they favor tax breaks?—an opinion that they either have or do not have. As a result, the nature of the problem is quite different.

A simple analogy will illustrate what is meant. Suppose we have a large box that contains thousands of red and blue marbles and we want to know what percentage of the marbles are red and what percentage are blue. The question is, Can we draw a sample of marbles out of the box in such a way that we draw very nearly the same percentage of red (or blue) marbles as exists in the whole box?

The answer is definitely yes. However, to be quite sure this happens, the sample must be large and we must select each marble *randomly*. Each time we draw one out of the box, all marbles should have an equal or at least known chance of being drawn. As long as there is a known chance of drawing any one of the marbles on any draw, we can apply the scientific principles of probability theory to the sample. The theory allows us to have great confidence that the true proportion we would find if we counted every marble in the box is within a few percentage points of the percentage of red marbles drawn in a large sample.

As stated, the reason we can use a sample to estimate the characteristics of all marbles in the box is because the sample we select is *random*. Of course, selecting and interviewing people is not the same as drawing marbles out of a box. To make accurate estimates based on a human sample, we try to meet four requirements:

- The sample is large enough to yield the desired level of precision.
- Everyone in the population has an equal (or known) chance of being selected for the sample.
- Questions are asked in ways that enable the people in the sample to respond willingly and accurately.
- The characteristics of people selected in the sampling process but who do not participate in the survey are similar to the characteristics of those who do.

These principles *make surveys work.* They are essential to understand and are therefore discussed in detail in Chapter 2. For now, the key point is that the closer a survey comes to meeting these four requirements, the more confidence we can place in the results.

Who Conducts Surveys?

Because meeting the requirements of an accurate survey is a reasonable task, more and more businesses, households, and individuals are being called on to take part in this kind of research. A survey in which you may have participated is the largest sample survey in the United States: It is conducted in conjunction with the U.S. decennial census, most recently in April 1990.

Every ten years the U.S. Bureau of the Census conducts a **census** in which it attempts to count, or enumerate, every person in the United States. At the same time, the Bureau asks a **sample** of roughly one in six households to answer more detailed questions on the so-called "long-form" questionnaire. The sample survey is designed to collect more detailed information than the census—about income, migration, employment, and other characteristics of people who live in the United States. The Bureau selects its very large sample for the long-form survey so that it can make estimates about the characteristics of people in nearly every neighborhood and community in the country.

Election eve presidential polls and the Census Bureau's one-in-six survey are only two of many sample surveys regularly conducted in the U.S. Other examples include

- the Federal government's Current Population Survey, a monthly survey of about sixty thousand households that provides statistics on income, employment, education, and other characteristics and is conducted between the decennial census years;

What Is a Census and Who Needs It?

A census, like the decennial census conducted by the U.S. Department of Commerce, is a one-by-one count of an entire population. For large populations numbering in the thousands or millions, censuses are unwieldy and enormous undertakings. The 1990 Census of Population and Housing, for example, required over 350,000 workers and about $2.6 billion to complete. And many critics charge that complete counts are unrealistic: "The United States Census has become an institution clinging to what many statisticians consider a myth: the idea that the Government can count the nation's population the way a child counts marbles—1, 2, 3 . . . 249,999,998, 249,999,999, 250,000,000. The population is too large, too mobile and too diverse to count that way" (Gleick 1990).

Occasionally, however, a census is the only way to get accurate information, especially when the population is so small that sampling part of it will not provide accurate estimates of the whole. An example is a 1979 census survey of beekeepers in Washington State. Questionnaires were sent to all one hundred or so registered beekeepers in the state in order to gather information about their needs and opinions. In this case, a sample survey of, say, 20 percent of the beekeepers would have yielded responses from only 15 to 18 percent—probably not enough from which to generalize about the population.

- television viewer surveys that provide information about how many people watch certain television programs;
- marketing research surveys on the entire range of consumer products, from food items to automobiles.

These are examples of large-scale, expensive survey research projects, many of which are conducted at the national level. While some of their results pertain to specific regions of the country and even to some larger states and cities, they typically cannot say much about smaller communities. However, it is now practical for local governments, nonprofit organizations, and small businesses to do their own scientifically valid surveys at low cost. Recent advances in mail and telephone survey techniques make these methods relatively inexpensive yet very reliable. In some cases, face-to-face surveys are also within the reach of people who need information about local communities. Before embarking on such a project, however, it is important to decide whether a survey is indeed "the right tool for the right job."

Is a Survey Appropriate for You?

We began this chapter with examples of local- and state-level surveys. In each case, the survey was an appropriate tool because it could provide estimates of population characteristics—the percentage of community residents who favor a tourism-based economic development strategy, for example.

Of course, not all information needs are best met by conducting surveys. In some cases, other methods are more useful. Four alternatives that should be considered are

- using existing or secondary data;
- conducting an in-depth case study;
- doing content analysis; and
- interviewing people who are not selected randomly.

Let's see how these nonsurvey methods can be used. First, **secondary data** are numbers that have already been collected for another purpose and transferred to a publication or computer-readable format. Sometimes secondary data come from surveys, for example, from the decennial census. Other times they are collected for administrative reasons, in the course of an organization's normal business. For instance, the U.S. Bureau of Labor Statistics publishes employment data that it collects to administer the unemployment insurance program. And large grocery stores regularly analyze cash register tapes (another kind of "secondary data") to determine patterns of consumer food purchases.

Secondary data are sufficient for meeting many information needs. For example, a nonprofit group in Washington, D.C., recently studied a grant program run by the Farmers' Home Administration (FmHA). Using FmHA administrative records, researchers examined whether the grant program actually benefited the people for whom it was designed. They did not have to conduct a survey because the data they needed were already available.

A second research method that should be considered is conducting an **in-depth analysis** of one case, such as an individual community or business. The purpose of a case study is to understand thoroughly all the factors related to the research question, but only for a single case. The results often generate ideas that can be explored in detail later. However, they cannot be generalized to any larger population, for example, to *all* communities or businesses.

Consider the economic development organization interested in whether tourism might be an effective job-creation strategy for nearby communities. To analyze this question, the organization might do a case study of a small town that has actively pursued tourism development in the past. By interviewing local residents, businesspeople, and community leaders, the organization can learn how and why the town adopted the strategy and what the consequences have been. The result will be a set of detailed insights into the particular circumstances surrounding that community.

The third research method that might be appropriate is **content analysis,** a systematic study of written or other types of communication. These might include government reports, the Yellow Pages, TV commercials, or newspapers. Content analysis is often used to study policy issues. A nonprofit organization in Nebraska recently conducted a study on state-level economic development policies in the Midwest. Using a variety of public documents—including formal policy statements, legislative enactments, gubernatorial initiatives, agency plans, and commission reports—staff members systematically analyzed how state development policies impacted small agricultural communities.

The fourth method that should be considered is **interviewing individuals or groups** who are selected for their particular characteristics rather than at random. Such individuals are typically chosen for interviews because they are knowledgeable about the research question. Groups, especially "focus groups," are often comprised of people who are similar in some important way, for example, income level. Group interviews are commonly used in political campaigns to test public reaction to advertising strategies. (Focus groups are discussed in detail in Chapter 3.)

Such interviews tend to be less structured than formal survey interviews, and often more exploratory. Like case studies, the results of interviews with people who are not selected randomly cannot be generalized but can be used to generate ideas for later study.

Each of the four information-gathering techniques discussed above is well-suited for addressing a particular type of question. However, if your goal is to find out what percentage of some population has a particular attribute or opinion, and the information is not available from secondary sources, then survey research is the *only* appropriate method.

Ethics and Accuracy in Surveys

Survey research carries with it an obligation to follow certain ethical norms. Any time you ask people to participate in a survey, it is your responsibility to respect both their privacy and the voluntary nature of their involvement. If you ignore this obligation, you violate respondents' trust in you and in surveyors who follow.

Practically speaking, ethical surveying means that you encourage people to respond but do not pressure them in an offensive way. Clearly, this requires a judgment call that each researcher must make based on the particular circumstances of his or her survey.

Ethical surveying also means that you do your absolute best to ensure confidentiality. This means that you release the results of your survey in the aggregate, that is, for the sample as a whole or for subgroups within the sample, rather than in such a way that an individual's responses might be identified.

Most surveys are not anonymous, which means the people working on the study *can* associate individual responses with particular people. In a face-to-face survey, for example, interviewers know exactly who answers each questionnaire. In a mail survey, a master list of names is usually maintained so that people who do not respond can be sent a follow-up letter. Confidentiality means you *can* associate responses with particular people but you *do not.*

We talk specifically about how to respect privacy and voluntary involvement in Chapter 8.

The point to which we often return in this book is that survey research is a powerful, scientific tool for gathering accurate and useful information. Its value comes from the idea of drawing probability (or random) samples from a larger population. It is also a tool that is frequently misused, often enough that the general public has good cause to be skeptical and very watchful. Many surveys that claim to be scientific yield inaccurate and misleading results, often because their samples are not random.

Consider, for example, the network TV news show that asked viewers to call in and "vote" on who won the 1992 presidential debates. Or think about U.S. senators and representatives who measured public opinion on gays in the military by how many calls came in on one side or the other. We have no reason to believe these are random samples from the whole population. In both cases, people *selected themselves* to register their opinion. Hence, we cannot draw inferences from their responses about how the general public feels. We hope this book will help people understand why some so-called surveys deserve skepticism and others produce useful information.

Types of Surveys

Before looking at the tasks involved in doing a successful survey, we return to three of the examples from the beginning of this chapter—one done by an economic development council, another by an art museum, and a third by a state employment agency. Each illustrates a reason why someone might want to conduct a survey.

- A **needs assessment survey** is used to solicit public opinion about community problems and possible solutions.
- A **marketing survey** is used to evaluate the nature and level of demand for particular products or services.
- An **evaluation survey** is used to learn about the impact of public or private programs and policies.

Each kind of survey is done to find out the characteristics, behavior, or opinions of a particular population. This is the underlying, primary objective of surveys described in this book. Whether the objective is achieved or not depends on the surveyors' success in undertaking 10 tasks, which we introduce in the next section and explore in the rest of the book.

In contrast, we will not specifically address another set of reasons for conducting surveys, namely, to organize communities, identify leaders, and create a network of people with similar interests. These are legitimate objectives for doing surveys, but they entail very different tasks than those described here. We hope it will become clear as the book progresses that using a survey to accomplish organizing and similar objectives may conflict with collecting information to describe a particular population, *especially if that information comes from a sample.* Hence, we strongly urge you to keep the information-versus-organizing distinction in mind as you read the remaining chapters.

Ten Steps for Success

A successful survey produces sound data that can be translated into valuable information for its intended users. Success is most likely if you pay attention to ten individual steps, each of which is the subject of one of the following chapters in this book.

1. Understand and avoid the four kinds of error. (Chapter 2)
2. Be specific about what new information you need and why. (Chapter 3)
3. Choose the survey method that works best for you. (Chapter 4)
4. Decide whether and how to sample. (Chapter 5)
5. Write good questions that will provide useful, accurate information. (Chapter 6)
6. Design and test a questionnaire that is easy and interesting to answer. (Chapter 7)
7. Put together the necessary mix of people, equipment, and supplies to carry out your survey in the necessary time frame. (Chapter 8)
8. Code, computerize, and analyze the data from your questionnaires. (Chapter 9)
9. Present your results in a way that informs your audience, verbally or in writing. (Chapter 10)
10. Maintain perspective while putting your plans into action. (Chapter 11)

Recently, someone who had never conducted a survey came to us for help. He planned to do a mail survey. Having heard that a 50 percent response rate was adequate for a valid survey, he wanted to know how to achieve that rate. We could have quickly answered his question with information from Chapter 8 about cover letters and follow-up contacts. Instead, we responded with a series of our own questions. "What do you hope to achieve by doing a survey? How large will your sample be? How much money can you afford to spend on data collection?"

At first our visitor was uncomfortable and asked more than once, "Why does that matter?" Our answer was one of the most important themes in this book: All the steps are related to each other, and each makes a critical difference to the success of a survey. It is important to understand the whole process before you can match your own resources with practical methods to produce useful results. Hence, the decision to conduct a survey should rest on whether you can confidently carry out the 10 steps described in the rest of this book.

For more details on nonsurvey research methods, see the following.

General

The Practice of Social Research, by Earl Babbie, Wadsworth Publishing Company, Belmont, CA, 1986.

Methods of Social Research, by Kenneth D. Bailey, The Free Press, New York, 1987.

Using secondary data

A Community Researcher's Guide to Rural Data, by Priscilla Salant, Island Press, Washington, DC, 1990.

"Role of Secondary Data," by Paul Voss, Steven Tordella, and David Brown, in *Needs Assessment: Theory and Methods,* edited by Donald Johnson, et al., Iowa State University Press, Ames, IA, 1987.

Content analysis

Basic Content Analysis, by Robert Phillip Weber, Sage University Papers Series, No. 07-049, Sage Publications, Newbury Park, CA, 1985.

Focus groups

Focus Groups as Qualitative Research, by David L. Morgan, Qualitative Research Methods, Vol. 16, Sage Publications, Newbury Park, CA, 1988.

Case studies

Case Study Research: Design and Methods, by Robert K. Yin, Sage Publications, Newbury Park, CA, 1989.

2

Cornerstones of a Quality Survey

The goal of this chapter is to explain why some survey results are more accurate than others. **Accuracy** has a special meaning in survey research: It describes results that are close to the true population value. For example, if data from a survey indicate that a county's per capita income is $12,529 and the true value is $12,671, we can say the survey result is relatively accurate. If another survey finds that 46 percent of U.S. voters support a new crime bill and the true value is 48 percent, that result is also relatively accurate. Sample surveying is the art and science of coming close!

A survey in a western state asked voters about their governor's economic development proposal. Results indicated that 51.4 percent of the voters supported what the governor had proposed. The staff person in charge of the survey was elated when he heard these results; the governor's position had won! In contrast, the researcher who directed the survey was horrified. The results she worked so diligently to produce had been interpreted like the outcome of an election, as if the "51.4 percent" was an exact measure that proved majority support.

At their very best, surveys can produce close *estimates* of what people think or do. In the case of the governor's poll, the results should have been reported as estimates—51.4 percent plus or minus a few percentage points. To attribute more exact meaning to the results is to misunderstand what makes surveys accurate or, in some cases, inaccurate. In this chapter we explain why some survey results are more accurate than others and how to improve accuracy within the limits of what is possible.

Errors That Affect Accuracy

Sample surveys yield accurate results when researchers succeed in avoiding four kinds of errors. To introduce the technical terms used to describe these errors, we look at the following four surveys, which did not produce accurate estimates, each for a different reason.

Case 1 In Las Vegas, Nevada, a crime prevention project used a telephone directory to obtain a sample of names and addresses for a mail survey. Unfortunately, the survey did not produce accurate (or useful) results because more than half of the residents of Las Vegas have unlisted telephone numbers. Furthermore, the population turnover is so great that many who have listed numbers are not in the most recent directory. This survey had substantial **coverage error**.

Case 2 A planner in a small community decided to interview a sample of residents about the attitudes they had toward the local school system. He developed a relatively complete list of the population from utility companies and voter registration records. He then randomly chose 50 residents for his survey. A sample size of 50 was not big enough to allow the planner to confidently generalize from the sample to the entire community. This survey had too much **sampling error**.

Case 3 A local council of ministries sponsored a survey of community residents to learn more about their participation in religious activities. Included in the survey was the following question: "How often do you attend church services?" The possible answers were:

1 REGULARLY
2 OCCASIONALLY
3 RARELY
4 NEVER

One long-time church member who attended services two or three times each month, answered "occasionally." Another person who attended services only at Christmas and Easter answered "regularly." The responses did not provide the council with useful information. This question suffers from **measurement error**.

Case 4 The president of a professional association decided to survey members about their journal publications. She printed the questionnaire on sheets of colored paper, which she then folded and mailed with bulk rate postage and no follow-up contacts. Only 15 percent of the members who received the questionnaire responded. This survey suffered from potential **nonresponse error**.

A perfectly accurate survey is seldom, if ever, conducted.* If it were, it would have these characteristics:

---

*Statisticians refer to sampling and some types of measurement error as *lack of precision*, and to other types of error as *lack of accuracy*. Together, all sources of error are referred to as **inaccuracy**.

1. Every member of the population that the researcher is trying to describe would have an equal (or known) chance of being selected for the sample. Hence, coverage error is avoided.
2. Enough people would be sampled randomly to achieve the needed level of precision. Hence, sampling error is minimized.
3. Clear, unambiguous questions would be asked so that respondents are both capable of and motivated to answer correctly. Hence, measurement error is avoided.
4. Everyone in the sample responds to the survey, *or* nonrespondents are similar to respondents on characteristics of interest in the study. Hence, nonresponse error is avoided.

Chances are that you have heard of sampling error before. For example, you may have seen the results of an opinion poll presented in the following way.

> Seventy-six percent of the 1422 adults polled by telephone from August 16th to 19th said they approved of the way [the President] was handling his job . . . ; only 15 percent disapproved. The margin of sampling error was plus or minus 3 percentage points. *(New York Times)*

Because some survey researchers tend to think most about sampling error, you may not have seen the additional caveat that the *Times* typically prints with its survey results: "In addition to sampling error, the practical difficulties of conducting any survey of public opinion may introduce other sources of error into the poll."

Virtually every survey exhibits these "practical difficulties," *in addition to* some level of sampling error. For example, some ideas and behaviors are hard to measure with a survey question, as in "Exactly how many books did you read during the last year?"

Another difficulty arises when it is impossible to obtain a complete list that gives everyone in the population a known chance of being surveyed. And of course, because surveys are usually voluntary, some people will not respond. As the proportion of refusals increases, so does the potential for nonresponse error.

Our key point is this: Anyone who does a survey should become accustomed to thinking in terms of *all four* sources of error, *at all times* during design and implementation. Although none of the four can be completely avoided, each has the power to render survey results useless.

We spend the rest of this chapter explaining the four types of errors in more detail. You will find specific recommendations on how to minimize them in each subsequent chapter.

Coverage Error

Coverage error occurs when the list—or frame—from which a sample is drawn does not include all elements of the population that researchers wish to study.

A **sampling frame** is a list of persons from which a sample is drawn. The author Robert Groves makes a distinction that is useful for understanding how an incomplete list can result in coverage error. He differentiates between the subset of people who are the focus of the research project—the target population—and the subset who are actually included in the sampling frame—the frame, or **survey population**. A discrepancy between these two populations results in coverage error.

The key point is that a sample survey can provide information about the population that makes up the frame *and no more*. Survey results do not pertain to people who are not on the list: We would not survey only residents of Omaha to find out what all Nebraskans think. Let's look at how the quality of the list affected the accuracy of one very famous sample survey.

In 1936 the *Literary Digest* conducted a mail survey to project the outcome of the upcoming presidential election between Alf Landon and Franklin Roosevelt. The *Digest*, a popular news magazine, had successfully conducted similar polls in previous elections. Using telephone directories and automobile registration lists to construct a sampling frame, the *Digest* sent ballots to a sample of some ten million people. Of this sample, two million people responded. Based on their responses, the *Digest* projected that Alf Landon would beat Franklin Roosevelt by 15 percentage points. The final election results in November 1936 were not as the *Digest* predicted. Roosevelt beat Landon with a record 61 percent of the popular vote, winning 523 electoral votes to Landon's 8.

Several serious shortcomings of the survey methodology were responsible for the difference between the *Digest*'s projections and the actual outcome of the election. One of these shortcomings was that the sampling frame consisted only of people who had telephones or registered automobiles, not of the entire electorate. The *Digest* could say nothing about the voting intentions of people who had neither telephones nor automobiles. These people were, by and large, poorer than those on the sampling list, and, as it turned out, they were far less likely to vote for Alf Landon.

The incomplete list introduced coverage error, of which the survey designers were unaware. Had the people who were excluded from the *Digest*'s sampling list been similar to those on the list in terms of presidential preferences, the coverage error would not have occurred. Unfortunately, they were quite different, and, as a result, the poll did not accurately predict the outcome of the election.

This example underscores a very practical problem with which all survey researchers must cope: Up-to-date, accurate sampling lists can be very hard to develop, especially for household mail surveys. In addition to being incomplete, lists sometimes contain duplicate entries, or entries that should be excluded because they aren't members of the target population. The level of coverage error depends on how different the missing, ineligible, or duplicate entries are from the target population.

We will describe specific techniques for overcoming coverage problems and for developing good lists in Chapter 5.

Sampling Error

Sampling error is a fact of life for those who conduct sample surveys. In other words, it can never be completely avoided unless you do a census. On the positive side, though, it is usually easy to quantify simply by referring to a table in a statistics textbook. (See, for example, Table 5.1 in this book.) Furthermore, sampling error is something the researcher can control just by increasing sample size. Because sampling error can be estimated, it is often the only specific error referred to when survey results are presented. The *New York Times* is an exception, which we noted earlier in this chapter.

> Sampling error occurs when researchers survey only a subset or sample of all people in the population instead of conducting a census.

To take the discussion one step further, we return to the marble example in Chapter 1. As before, suppose we have a box containing red and blue marbles. *Without counting every marble in the box*, we want to know what percentage of the marbles are red and what percentage are blue. As long as we pick the marbles randomly—that is, as long as there is an equal chance of drawing any marble on any draw—we can use a sample to estimate the number of red and blue marbles in the whole box. But because we draw a sample instead of doing a complete count, we can give only a ballpark estimate of the proportion of red and blue marbles. The dimensions of the ballpark (in other words, the sampling error) are directly related to the size and uniformity of the population. In fact, statisticians have spelled out how these variables are related to each other in a formula, which we present and explain in a technical note at the end of Chapter 5.

Measurement Error

> Measurement error occurs when a respondent's answer to a given question is inaccurate, imprecise, or cannot be compared in any useful way to other respondents' answers.

Measurement errors are made when we collect data, not when we devise lists or select samples. Let's assume for now that each item on a questionnaire has a "correct" answer. (This isn't always true, as explained in Chapter 6.) The size of the measurement error is the difference between a respondent's answer to a particular question and the "correct" answer. Measurement errors come from four sources—the survey method, the questionnaire, the interviewer, and the respondent.

The Survey Method. As described in Chapter 4, mail, telephone, and face-to-face surveys place very different demands on the respondent. In a mail survey, respondents control question pace and sequence. They can read ahead to get a picture of the overall context of the questionnaire as well as to find out how many more questions they will be asked. No interviewer is present to influence particular answers.

In a telephone or face-to-face survey, interviewers control the pace and sequence. Respondents must rely on what they hear when formulating their answers; the only context comes from what they remember of previous questions. Often people are influenced by what they think the interviewer would consider an "acceptable" answer.

Hence, it isn't surprising that the survey method affects how people respond. Researchers have found that the same question asked by mail, telephone, and face-to-face sometimes yields very different answers (Tarnai and Dillman 1992). This **method effect**, as it is called, is complicated. We cannot say categorically that one method produces more measurement error than another or in one direction or another. However, it is clear that measurement error is a more significant problem on some types of questions, such as those that deal with abstract ideas or sensitive issues, depending on which method is used.

We offer suggestions on how to minimize method effect in Chapters 6 and 7.

The Questionnaire. The measurement error example we cited at the beginning of this chapter concerned a survey of religious activities. Respondents were asked whether they regularly, occasionally, rarely, or never attended church. An error occurred because the answer choices were not clearly defined. The words were not understood in the same way by all respondents. Hence, the answers they gave were not comparable and the results were useless.

The questionnaire is the source of other measurement problems as well. For example, question *structure* may confuse the respondent, thereby making it impossible for him or her to answer "correctly." Consider the question "How did you first hear about the proposed sales tax change?" with the answer choices:

1 FROM A FRIEND OR RELATIVE
2 AT A MEETING OF AN ORGANIZATION TO WHICH I BELONG
3 AT WORK
4 FROM MY SPOUSE
5 FROM THE TELEVISION, RADIO, OR NEWSPAPER

The answers are not mutually exclusive. A respondent may have heard about the tax change, for example, from a friend at work or a relative who read about it in the newspaper. Hence, for this respondent, the question has no single "correct" answer. Some respondents who can logically pick two answers might pick both, while others might pick only one.

The Interviewer. In face-to-face and telephone surveys, the person who asks the questions is critical to data quality because he or she administers the questionnaire. Even with good training, interviewers have the opportunity to make many mistakes. For example, the interviewer may "lead" respondents by suggesting particular answers, as in:

The President recently increased the number of U.S. troops in the Persian Gulf. *You support this action, don't you?*

Alternatively, he or she may change the meaning of a question by rewording it in response to a respondent's inquiry, as in:

Interviewer: The President recently increased the number of U.S. troops in the Persian Gulf. Do you or do you not support this action?

Respondent: By "support," do you mean would I go to the Gulf or do I think it's a good idea?

Interviewer: I mean would you go.

Respondent: No.

The interviewer's reinterpretation is very likely to affect how the question is answered. Hence, it is impossible to compare answers across all respondents.

Finally, interviewers may bias respondents simply by the image they project. For example, arriving at the interview in a car with the bumper sticker "Support Our Troops in the Gulf!" is likely to affect how respondents think they *should* answer. If they perceive the interviewer as being neutral, they may be more willing to answer honestly.

The Respondent. In the final analysis, we cannot get "correct" answers to survey questions unless respondents are willing and able to provide them. Respondents can deliberately or inadvertently answer incorrectly, as in:

Interviewer: How much money did you receive from wages or salary in the last calendar year?

Respondent A: [Thinking to himself that the question is none of the interviewer's business, and answering] $463,229.

Respondent B: [Misunderstanding the question and answering for only one month instead of a calendar year] $1250.

Clearly, neither answer is correct.

We have seen that measurement error can occur at any time during data collection. To guard against it, try to choose the most appropriate survey method; write clear, unambiguous questions that people can and want to answer; and train your interviewers carefully. We address these issues further in Chapters 4, 6, and 8.

Nonresponse Error

Nonresponse error occurs when a significant number of people in the survey sample do not respond to the questionnaire *and* are different from those who do in a way that is important to the study.

Even if researchers compile a complete list or frame with no duplicates, draw a large enough sample, *and* make accurate measurements, they still have to contend with nonresponse error. This error is a problem if the following two conditions hold true at the same time.

- More than a small number of people who were selected in the sample are not interviewed, either because they cannot be reached or refuse to participate.
- Nonrespondents are different from respondents in a way that pertains to the study focus.

Consider a survey designed to identify which local businesses should be targeted to receive technical assistance from a new business development center. The researchers began with a list of businesses that filed tax returns the previous year, a list which they received from the state department of revenue. By contacting local utility companies, they supplemented this list with names of businesses that had been started since the end of last year. Next, they drew a random sample for purposes of conducting a mail survey. They mailed out the questionnaires and, after a follow-up letter, decided to stop when 40 percent of the sample had responded.

After the researchers coded and tabulated the data, they learned that 60 percent of the businesses who wanted technical assistance were large operations that sold products outside the community or state. They concluded that the new business development center should concentrate on helping these larger operations.

Unfortunately, the researchers made an error when they failed to consider the characteristics of businesses that were in the sample but failed to respond to the questionnaire. As it happened, these businesses were overwhelmingly smaller, more locally oriented and, in fact, wanted technical assistance but could not garner staff time to participate in the survey. Had they responded, the researchers may well have concluded that small businesses that sell locally should receive assistance from the development center.

In this case, the researchers should have seen the low response rate as a warning signal for problems with nonresponse error. For example, they might have looked at the relative sizes of the firms that responded. When they realized that small firms were underrepresented, they could have tried harder to get responses from this important part of the sample.

However, it is important to mention that getting a high response rate overall is not the same as avoiding nonresponse error. In 1990 over 98 percent of the population responded to the U.S. Census, but there was a significant undercount in certain regions of the country. Despite the high response rate overall, nonresponse error was a real concern in areas where the undercount occurred.

The first line of defense against nonresponse error is to aim for the highest response rate possible, a topic we deal with in Chapter 8.

When Is There "Too Much" Error?

Having read our explanation of the four error sources, you are justified in asking how much error is too much. For the most part, it is a matter of judgment and individual circumstances. Throughout the rest of this book, we suggest numerous ways to reduce error. For now, here are some general guidelines.

First, make every reasonable effort to minimize coverage error by using or compiling the best sampling frame. (See Chapter 5 for particulars.) Still, coverage error will probably remain a problem to some extent. It cannot be measured directly, but you can think about how it might affect the results. Consider who might have been excluded from the survey and how they might differ with respect to characteristics important to the study.

Second, the level of sampling error *can* be controlled by selecting a smaller or larger sample. (See Chapter 5 for specifics.) The hard decision is how much error you can tolerate. Will it matter if your estimates are within 5 percent of the actual value, or is more precision required? If dividing up a large Federal grant depends on estimates derived from the survey and you want to make sure the division is fair, by all means minimize sampling error. But if the goal is to get a general idea about the number of local businesses that are considering expansion in the next five years, perhaps you can tolerate less precision.

Third, avoid obviously biased or vague questions and other areas where measurement error might creep into the results. (See Chapters 6 and 8 for details.) One usually cannot say with any confidence how large the measurement error is, but it can be minimized with careful question wording, pretesting, and thorough interviewer training in telephone and face-to-face surveys.

Finally, design a questionnaire and implement the survey so as to get the highest possible response rate. In this way, the chances of nonresponse error will be minimized. A low response rate serves as a warning that nonresponse error might be a problem. Depending on who is surveyed and what method is used, anything under 60–70 percent should be a red flag—roughly 60 percent for a general-public mail survey, about 70 percent for a special-population telephone survey. If the flag goes up, get out your magnifying glass and detective hat. Find out whether the people who didn't respond are different from those who did in ways that matter to the study. Does the mailing list indicate that the two groups are different in terms of geographical location? For example, are they from different neighborhoods or counties? Or if you survey members of an organization, how do respondents compare with others in terms of length of membership?

When prior knowledge like location or length of membership is unavailable, compare respondents to the larger population by using data that have already been collected by someone else. In a household survey, compare demographic characteristics from the census of population. In a business survey, compare establishment characteristics from the various economic censuses. (Data from both types of censuses are published by the U.S. Department of Commerce.) Look for clues about how well survey respondents represent the population that interests you.

So an answer to the question about how much error is *too much* is not straightforward. All we can say is this: Be aware of the four sources, minimize them as much as reasonably possible, and use good judgment.

Summary

According to one old saying, if the only item in a person's toolbox is a hammer, it shouldn't be surprising if everything around him looks like a nail. In many ways, this piece of folk wisdom describes the dilemma addressed here. Some people tend to assess the confidence they should have in survey results only as a function of response rate. Others focus on the sample frame and whether everyone in the particular population had an equal chance of being selected. Still others zero in on whether questions are worded in such a way that respondents can answer unequivocally. And finally, some people simply look at the absolute number of people who participated in the survey, regardless of the response rate.

Just as a toolbox with only a hammer is inadequate, so is the evaluation of a survey that focuses on just one cornerstone of survey quality. Resisting the temptation to favor one over another is hard to do. Media reports often discuss just one—usually, the response rate—and even survey methodologists often have a favorite; typically, it is the one that their training equips them to understand best and do something about.

Colleges and universities do not teach one single discipline called "survey research." Instead, relevant courses are scattered across the curricula of many fields. With their grounding in mathematical theories and concepts, students trained as statisticians typically deal best with sampling error and, to some extent, with coverage error. In contrast, psychologists and sociologists are more oriented toward human behavior. Hence, they focus on measurement error and how it is influenced by respondents, interviewers, and question wording. They are also concerned with nonresponse error.

The critical importance of this chapter is to establish that your survey is subject to four, very different sources of error, all of which deserve attention. Methods of controlling these four sources of error make up your toolbox. Use it often as you proceed through your survey, the steps of which we describe in subsequent chapters.

For more detail on the four error sources, see the following.

Chapter One, "An Introduction to Survey Errors," *Survey Errors and Survey Costs*, by Robert M. Groves. Wiley Interscience, New York, 1989.

3

Deciding What Information You Need

From this point forward in the book, we assume you have decided that conducting a survey is the best way to get the information you need. Before getting embroiled in the mechanics of conducting a survey—such as sample selection and questionnaire design—it is important to *specifically* answer the question of why you need new information in the first place.

Throughout this book, we encourage you to keep in mind the four types of error. We stress over and over that surveys fail when errors are ignored and data are inaccurate. In this chapter, we address another, equally important problem—the failure to define survey objectives.

Why is defining the purpose of a survey so important? Because no amount of money or talent can create value out of a trivial question. Even the most accurate data cannot redeem an irrelevant survey. If the study addresses an issue that doesn't matter to anyone or if the questions don't ask for useful information, the project will be a waste of time, money, and energy.

Be Specific

Making sure the survey will provide useful information means raising two specific questions: What *problem* are you trying to solve? and What *new information* do you need to solve it?

The *problem* is the fundamental issue you are trying to address with your survey. It may be high unemployment, falling attendance at church, inadequate waste disposal, or a host of other difficulties that communities and businesses face. The key is to be specific about what the problem is, why you believe it is important, and what you already know about it.

Presumably, you are doing a survey to get *new information* that will help you solve a problem, so it is important to be precise about what you actually need to know. At first, the questions you pose may fall short—they may be *vague, biased,* and/or *not really critical* to solving the problem.

Vagueness: We have often found ourselves working with someone at the beginning of a survey who insists repeatedly, "I want to know people's attitudes about . . . ." The attitudes may be toward lawyers, tourists, or big business, or about any of a number of issues. They may even ask if we have questions concerning attitudes from another survey so that they could use someone else's wording. It is as if they were shopping, didn't know what they wanted to buy, but might recognize it if they saw something similar. People whose objectives are this vague need to better define what they want to know.

Consider the college administrator trying to reverse a recent increase in the dropout rate among freshmen. Her first attempt at telling us what information she needed was, "I want to know freshmen's attitudes toward our school." After several attempts, she was able to explain more specifically what she wanted to know. For example, she wanted to know whether students thought they had adequate financial resources to get through school, whether they were satisfied with classroom instruction, whether professors were available for individual help, and whether residence halls were too crowded. These clear descriptions of what she needed to know helped us develop a list of questions for her survey.

Bias: We've also been asked to do surveys to provide information that would prove a point. As a county commissioner once told us, "I want to show that people in this county agree with our decision to build a recycling facility." In other words, this person wanted information that would support what he already "knew" was true.

Indeed, we could have done a survey to estimate the percentage of residents who wanted to build a recycling center. We also could have used the survey to learn who most strongly supported the facility and who thought it was a waste of money. However, designing a survey to *prove* the commissioner's point would have openly invited error. People who already know the answer to their questions almost always make mistakes that lead to biased results. They write leading questions, pick respondents who are sympathetic to their position, and ignore results with which they disagree.

When someone has preconceived ideas about what a survey "should" show, we encourage them to step back from their expectations and reevaluate whether they really need new information at all. It may be that the survey won't help them solve the problem they are trying to address.

Noncritical: Finally, when talking to people about their information needs, we often end up with lengthy lists of interesting things to know. One such brainstorming session produced 200 separate questions that a citizens' group wanted to answer about their community. Without further thought, they proceeded to put all 200 interesting (but nonessential) items into a questionnaire and mailed it out. The lucky recipients asked one another, "Did you get that laundry list yet!?"

To everyone considering a survey, we suggest, first and foremost, that you articulate your problem and decide what information you need to solve it. Putting your objectives on paper will clarify your purpose and will provide a document for potential funders and other people involved in the project.

In the next section, we suggest an exercise to help you define your objectives clearly and narrow your focus.

Think in Terms of Results

Surveys are used to estimate the characteristics, behaviors, or opinions of particular populations. Sometimes, the only reason a survey is done is to get one or two numbers for a grant proposal. For example, we may need to know the percentage of community residents with incomes below the poverty level or the vacancy rate in local housing units.

Usually, however, we want more from surveys. In addition to knowing how many people are poor, we want to know *why* they are poor. In addition to knowing how many housing units are vacant, we want to know *where* the vacancies are and in *what kind* of units. To zero in on details like these, think in concrete terms about the numbers your survey will produce. A good way of doing so is to prepare hypothetical tables of results.

Consider the example of a school superintendent who wanted to estimate voter support for a proposed bond issue that would finance construction of a new elementary school. In an earlier election, voters rejected the bond, yet the superintendent was convinced a new school was necessary. Her problem was getting the necessary two-thirds of the community to agree with her position. She decided to do a survey to learn how voters felt one month after the first election. Specifically, she wanted to know (1) whether voter opinion varied according to neighborhood and (2) why some people had voted against the bond. She planned to use this information to run a better campaign and then hold another election.

At our first meeting with the superintendent, we asked her to speculate about different ways the survey might turn out. Then, using the possible outcomes she proposed, we prepared two tables with hypothetical results, as shown in Figure 3.1.

Numbers in the two tables—the survey *results*—would give the superintendent information she needed to focus on one particular campaign strategy. For example, survey results like those in Table A, Column A (of Figure 3.1) show very uneven support that varies according to where people live. In some neighborhoods, support was high enough to get the bond passed, whereas in others voters were strongly opposed. These results would tell the superintendent she needed to target her campaign on particular neighborhoods. In contrast, results in Column B would indicate little variation

Figure 3.1
Preparing hypothetical results helps separate "need to know" from "nice to know."

Table A. What the superintendent's survey might show about whether voter support for the school bond issue varies by where people live

	Percent in favor (Column A)	Percent in favor (Column B)	Percent in favor (Column C)
Southport neighborhood	40	40	68
Carriage Hill neighborhood	20	38	63
College neighborhood	70	42	70
Pioneer neighborhood	70	46	68
Outside city limits	60	48	62
All neighborhoods together	60	46	66

Table B. What a survey might show about why the bond did not pass in the first election

	Percent of voters who think this is an important reason why the bond was defeated	
	(Column A)	(Column B)
Other schools should be expanded before another school is built.	10	45
Neither improvements nor another school is needed in this community.	15	20
The bond would have raised taxes beyond what people can afford.	56	15
People don't trust decisions made by the superintendent and school board.	20	60
People don't understand what the bond issue was about.	21	33

by residence but, overall, only lukewarm support for the bond. That would mean a big, community-wide campaign to win the election. Results in Column C would suggest that the bond could probably pass with fairly even support across the community, given a moderate amount of successful campaigning.

Survey results like those in Table B (of Figure 3.1) would tell the superintendent which issues were most important to address in the second campaign. The percentages in Column A would suggest that people simply could not accept the price tag—somehow, the cost of the proposed new school would have to be scaled back. In contrast, the percentages in Column B would tell the superintendent that she had serious public relations problems. People would be willing to pay for the new school but only if they had greater trust in administrators.

Preparing hypothetical tables forces us to separate "need to know" from "nice to know." If the school superintendent had no way to target her campaign *by neighborhood,* the kind of information in Table A would simply be "nice to know." It wouldn't help her win the second election, and she therefore should probably leave questions about residential location off the survey. "Need to know" is a critical criterion for every item on a questionnaire.

Focus Groups Can Help

An important part of ensuring that the survey will provide useful information is to involve the intended audience in answering the two questions we posed above: What problem are you trying to solve? and What new information do you need to solve it? A good way to do this is to conduct informal, loosely structured interviews. Depending on the nature of the survey and the preferences of people who are concerned with the project, you may decide to interview people individually or in a group. In either case, seek ideas from those who will use the results—for example, city council members or health care providers—and from those about whom the survey is concerned—for example, elderly or unemployed members of your community.

In addition to providing insights about the subject of the survey, such interviews offer another advantage. They encourage people other than those doing the surveys to "buy in" and support the research. People who have brainstormed about the survey are more likely to accept the results as being reliable, and ultimately, to use them.

One technique that can be used in this planning phase is the "directed group discussion" or focus group. Focus groups are organized discussions led by a moderator; they typically involve 8 to 10 people. The purpose of a focus group is to stimulate people's thinking and elicit ideas about a specific topic. For instance, focus groups are used by private businesses to learn how customers respond to new products and by political campaigns to test voter opinion. They have also been used effectively by public agencies as a means of identifying the range of perceptions and ideas about a program or service. For example, agricultural cooperative extension services have used focus groups to evaluate potential course offerings and qualities of successful 4-H programs.

Focus groups do not substitute for quantitative surveys because the participants are not randomly selected nor do they comprise a sufficiently large sample to yield reliable estimates. Therefore, they *cannot* reveal what proportion of a population has a particular attribute or opinion. For example, we cannot conclude from a focus group whether the majority of residents in a community are willing to pay higher gasoline taxes.

Focus groups *can,* however, provide a head start on knowing which questions to ask in a survey. For instance, they can suggest the terms under which residents might be willing to pay higher gas taxes: in return for more traffic lights downtown, better snow removal, or new sidewalks. Surveyors can then use such information as a guide to developing their questionnaire. (We'll see in Chapter 6 that it is usually better to offer people a series of responses than to leave questions completely open-ended.)

At the outset, we want to emphasize that opening up the planning process through public participation does not mean compromising the scientific integrity of a survey. In our experience, organizers of the most successful surveys manage to achieve a delicate balance between principles of sound, unbiased research on one hand and community involvement on the other. Short measures of either one translate quickly into final reports that gather dust on office shelves.

Focus Groups Cannot Substitute for Surveys

	Focus Groups	Surveys
Purpose	To stimulate thinking and elicit ideas on a particular subject	To determine what proportion of a predefined population has a particular attribute or opinion
Structure	Discussion of a small group of people, led by a moderator	Mail, telephone, or face-to-face questionnaire, completed by an individual respondent
Capacity to generalize to a larger population	No	Yes
Capacity to generate ideas or hypotheses for later testing	Yes	To some extent
Capacity to test ideas or hypotheses	To some extent	Yes
Must questions and answers be formulated ahead of time?	No, but the moderator must be ready to guide the discussion	Yes, except for open-ended questions

Summary

In this chapter, we've suggested that you be specific about the problem you're trying to solve and the new information you need. We've also shown how thinking in terms of results can help narrow your focus. Finally, we've offered some ideas on how to get public input on survey planning.

If defining objectives is such a critical part of doing survey research, why is this chapter so short? The answer is that, despite its importance, there is no magical formula for making sure a survey serves a useful purpose. We can provide only guidelines and encourage you to be as diligent at this early stage as you are when you actually start collecting data.

Next we turn to a question with serious implications for your bank account—Which survey method is best for you?

4

Choosing a Survey Method

In this chapter, we discuss the choice between mail, face-to-face, and telephone survey methods. No single method can be judged superior to the others *in the abstract*. Instead, each should be evaluated in terms of a specific study topic and population, as well as budget, staff, and time constraints.

Before explaining the considerations that go into choosing a particular method, let's look briefly at how each one works. (Details are in Chapter 8.)

- **Mail surveys.** Surveyors select their sample from a reasonably complete address list of a population. Next, they may mail an advance-notice letter, followed by a questionnaire (with a cover letter and stamped return envelope), to each member of the sample. Then, within about a week, they send a postcard reminder. Respondents complete the questionnaire and mail it back. People who do not return the questionnaire promptly can be contacted again, either by mail or telephone (if phone numbers are available).
- **Telephone interviews.** Surveyors select their sample from a telephone directory or other list. Alternatively, they can use one of several random-number techniques that have been developed to overcome list problems for general-public surveys. People in the sample are interviewed at the time of the first phone call or at another, more convenient time. Interviewers record answers either on a survey form or, for some surveys, directly into a computer.
- **Face-to-face interviews.** If either a telephone or address list is available (or can be compiled), surveyors select their sample from the list and then contact each member to conduct the interview in person. If no suitable list is available, an area frame sampling technique is used. Interviewers conduct the survey by talking to respondents in person and recording the answer to each question on a survey form. If the respondent is not home when the interviewer arrives or if the interview is interrupted, he or she must make one or more additional visits to complete the job.

A fourth method, the **drop-off survey,** combines features of face-to-face interviews with mail surveys. In this case, surveyors go door-to-door, personally delivering questionnaires to individual households or businesses. Respondents complete the questionnaires on their own and then either return them by mail or keep them for the surveyors to collect.

Choosing a method is sometimes quite easy. Consider the case of a chamber of commerce that wanted to assess a range of service activities that it might undertake in the future. At the annual meeting, members voted unanimously to conduct a chamber-wide survey to gather the necessary information. They decided to do a mail survey, using a current address list of all member businesses and their owners. The questionnaires were mailed, completed, and returned promptly to the survey director.

Next, consider a community action agency whose staff was preparing a grant application to the Federal government. The agency was applying for funds to set up a day care project and had only two weeks to get the proposal in. Much to their alarm, staff members discovered that to complete the proposal they needed to estimate how many local preschool-age children lived in households in which all adults worked outside the home. Because they needed the information so quickly, they decided to conduct a telephone survey.

Finally, consider a city council that sought a solution to a worsening problem of homeless people. Council members had several options for providing shelter to the homeless, including a vacant school building, city park campsites, and downtown boarding houses. They decided to survey a sample of homeless people about which location was most acceptable and likely to be used. The survey involved face-to-face interviews conducted on street corners, at soup kitchens, and in the bowery.

These three surveys focused on dramatically different populations. The chamber of commerce surveyed its own members, who understood the intentions, wanted to respond to the questionnaire, and for whom the chamber had an accurate list. The community action agency surveyed households who were very likely to have telephones and to welcome efforts to improve local day care services. The city council had no list of homeless people and, furthermore, was likely to encounter respondents who were not interested in or capable of completing a written questionnaire.

Choosing a survey method for these three projects was unusually straightforward. It was clear that a mail survey would work for the chamber, a telephone survey was appropriate for the action agency, and only face-to-face interviews were suitable for the city council. In most cases, other considerations mean the decision is not as clear-cut. We spend the rest of the chapter explaining what these considerations are and how they pertain to each method.

What to Consider

The first thing to think about when choosing a survey method is the kind of resources you can commit to your project, including

- how many people are available to work on the survey, either volunteer or paid, and whether they have survey experience;
- how much time you have to produce results;
- whether someone with survey experience can help you and at what price;
- what facilities you have at your disposal, especially with respect to telephones; and finally,
- how much money you can spend on the survey.

Most of this chapter is about how these resources play into the choice of a survey method. Resources—especially money—aren't the only thing to consider though. Another important consideration is how sensitive each method is to various kinds of errors. Recall that in Chapter 2 we introduced four error sources: coverage, sampling, measurement, and nonresponse. Mail, telephone, and face-to-face surveys are each sensitive to these four error sources in varying degrees. For example, the *Literary Digest* survey described in Chapter 2 illustrated how coverage error from an incomplete list can bias the results of mail surveys—so much, in fact, that the data may be rendered useless.

Each of the four error sources can be a potential problem at different stages of a survey. Each can be more or less troublesome depending on what resources are available. For example, consider choosing between the telephone and mail methods. Telephone surveys are very sensitive to errors caused by interviewers who don't read questions exactly as they are worded in the questionnaire. In contrast, mail surveys don't involve interviewers and, therefore, are not sensitive to this particular type of measurement error. All else being equal, a telephone survey is more appropriate when trained interviewers are available, and a mail survey is better when they are not.

Hence, which method is best for a particular survey depends not only on budget, staff, and time constraints, but also on which kinds of error are likely to be encountered. With these things in mind, we describe each method in detail in the next section.

Mail Surveys

The greatest strength of mail surveys is that they require the least amount of resources. Respondents—not interviewers—fill out the questionnaires, and fewer people are thus required to conduct the survey. The skills needed are primarily clerical: typing, sorting, and processing correspondence.

In addition, mail surveys are the easiest to do for people who have no experience and no professional help. One can spend weeks or even months designing the questionnaire and preparing to make follow-up contacts. But when the survey starts, there is little to do besides crossing names off the mailing list, processing incoming questionnaires, and preparing the next mailing. In contrast to telephone and face-to-face surveys, mail surveys do not require decision making on an immediate, high-pressure basis.

Another strength of mail surveys is that they allow one to minimize sampling error at relatively low cost. Because of lower staff requirements, the extra cost of sending out and processing more mail questionnaires is less than conducting additional telephone or face-to-face interviews. Therefore, people on a tight budget may be less inclined to cut costs by decreasing sample size, which tends to increase sampling error.

Mail surveys, especially when a survey is to be done locally, can provide a sense of privacy. It is easier for most people to answer personal questions in writing than face-to-face, with an interviewer they may actually know. The mail questionnaire is a more anonymous vehicle for giving information about income, mental health, political attitudes, and a host of other issues people often consider private.

A final strength of mail surveys is that they are less sensitive to biases introduced by interviewers as well as to the tendency for respondents to give answers they think the interviewer wants to hear. As we'll see below, these kinds of measurement errors are potentially very serious in both telephone and face-to-face surveys.

The greatest weakness of mail surveys stems from their sensitivity to noncoverage error, as the *Literary Digest* example in Chapter 2 illustrates. Published lists, from which samples are often drawn for mail surveys, are almost never complete. For example, lists of telephone subscribers, utility customers, or car owners are typically incomplete for a variety of reasons: Some members of the population lack telephones or utilities; others keep their names off the lists to maintain privacy; and, finally, the lists are often out-of-date. Even when such lists are relatively complete, they may be confidential and not available to someone doing a survey.

A second weakness of mail surveys is that some people are less likely to respond to the questionnaire than others. Hence, nonresponse error can be a serious problem. People who receive a mail questionnaire have the chance to examine it before deciding to respond; their interest in the topic will very likely affect this decision. For example, in a mail survey about purchases of prepackaged fruit and vegetables, people who are concerned about the environment may be more likely to respond. Those who are less concerned about the environment—and more likely to buy prepackaged produce—may consider the survey a waste of time. Respondents differ from nonrespondents in a way that affects the survey results.

Those who cannot read the questionnaire, follow its instructions, or provide written answers are unlikely to respond to a mail survey. If these nonrespondents are less educated or older than respondents, for example, the survey results may be biased. But testing for this bias in mail surveys is often extremely difficult. We usually have no way of knowing the characteristics of nonrespondents without making a personal contact.

Another weakness of mail surveys is that researchers have little control over what happens to the questionnaire after it is mailed. They cannot be sure the correct person in the household or business fills out the form or whether the intended respondent receives advice from others in answering the questions. For example, even though a questionnaire clearly states that the owner of a business should answer the questions, he or she may tell an employee to do the job. And even though the instructions may state that survey sponsors are interested in the opinions of the oldest member of the household, he or she may ask someone else for their opinion. These are important problems because we have no way of knowing when they occur and so are helpless to prevent them.

Surveyors also cannot control whether mail questionnaires are filled out completely. Respondents may purposely skip over difficult and boring questions, or inadvertently overlook some items. Both cases of **item nonresponse** are easier to avoid in telephone and face-to-face surveys.

In summary, mail surveys are best suited for

- surveying people for whom a reliable address list is available and who are likely to respond accurately and completely in writing;
- surveys in which an immediate turnaround is not required; and
- projects in which money, qualified staff, and professional help are all relatively scarce.

Telephone Surveys

People who have decided to do a survey are often reluctant to consider using the telephone method. They think about their own reaction to unsolicited telephone calls, and they worry that people will not respond.

Indeed, refusals are very common in national, general-public surveys. However, in small-scale surveys like those of interest to most of our readers, nonresponse tends to be a less important problem. For example, parents of children in the county 4-H program are unlikely to hang up on an interviewer who wants to survey them about the program's popularity. Local library patrons are unlikely to hang up on an interviewer asking questions about proposed library fees. Hence, when potential respondents are apt to be familiar with the issue being studied, the telephone method offers real advantages.

The greatest strength of this method is its ability to produce results quickly. Companies like Gallup use telephones to conduct public opinion polls during a one- or two-day period and report results almost immediately. Smaller survey organizations also take advantage of the rapid turnaround offered by telephone surveys.

There are several reasons why telephone surveys can produce results more quickly than other methods. First, interviewers who use a telephone can complete more interviews in a given time period than those who must physically travel to someone's house or business. A good telephone interviewer can complete and check more than three 30-minute interviews during a three-hour calling period, but the same person doing face-to-face interviewing might only be able to complete one.

Second, if the survey is conducted at a central facility equipped with a bank of telephones, a supervisor can deal immediately with any problems that arise. If a particular question in the survey causes problems or if a respondent wants to talk to someone other than the interviewer, the supervisor can respond quickly. The same problems occurring in a mail or face-to-face survey can delay the process by days or even weeks.

In addition to quick turnaround, telephone surveys offer the advantage of greater interviewer control. In contrast to mail surveys, phone interviewers can ask to speak with the person they want to answer the questionnaire, encourage the respondent to answer all the questions, and avoid the influence of others in the household or business.

The cost of telephone surveys lies generally between that of face-to-face and mail surveys; its two main components are labor and long-distance charges (if the calls are not local). Face-to-face surveys have higher labor costs than telephone surveys because fewer interviews can be completed in a given time period. Mail surveys have lower labor costs because respondents, rather than interviewers, fill out the questionnaires. They also don't entail long-distance telephone charges.

Telephone surveys are not without weaknesses, of course. One is that not all people have telephones. Hence, a subgroup of the population is automatically excluded from being surveyed. Since about 93 percent of all people in the United States live in households with telephones, this is not a serious problem for nationwide surveys of the general public. However, it is a major drawback for surveying certain groups of people. Those who live in the South and in rural areas are less likely to have telephones than the general public, as are those who have not completed high school, are black, have low income, live in large households, or are unemployed. For example, about 24 percent of Southerners who have less than a high school education have no telephone; the same is true for 27 percent of all people in the U.S. whose income is below the poverty level (Thornberry and Massey 1988).

Another reason some people are deterred from conducting telephone surveys is that telephone directories—the easiest lists from which to draw samples—are incomplete. About one in five households in the U.S. moves every year, so directories are inevitably out-of-date. In addition, some households have unlisted numbers (although fewer in rural than urban areas). And increasingly, more households have more than one listing—one for each spouse, for example. Each of these situations presents a problem to the surveyor who would ideally like every member of the population to have an equal (or known) chance of being selected in the sample.

Fortunately, several techniques have been developed to overcome incomplete and inaccurate directory problems. Both **random digit dialing** and **add-a-digit dialing** make it possible to access both listed and unlisted numbers; these techniques are explained in Chapter 5.

Other problems with telephone surveys can be harder to overcome. The first is that a knowledgeable supervisor is critical to success, especially when interviewers have not conducted surveys before. Everything must come together efficiently and quickly when interviewing begins. Inevitably, a host of "What do I do now?" questions come up when the first calls are made. For example:

> A new family lives in one of the houses in our sample. Do I interview them or track down the original residents?
>
> The respondent only has time to talk to us at 6:30 tomorrow morning. Will someone be here to conduct the interview?

> We're supposed to interview renters in this survey. But these people do maintenance work instead of paying cash for their rent. Do they qualify?

A telephone survey cannot produce accurate results unless a knowledgeable supervisor is on hand to answer these kinds of questions.

Other problems with telephone surveys have to do with their sensitivity to measurement error. Telephone interviews depend completely on what can be communicated vocally. To understand what is being asked, the respondent must concentrate on each word or phrase. Questions in which he or she is asked to rank a series of items are very difficult to use over the phone. The same is true of questions that depend on maps or diagrams. Compounding the problem is that interviewers cannot observe respondents' reactions for clues as to whether questions are understood.

Finally, respondents in telephone surveys can easily be influenced by leading questions from the interviewer ("Don't you think that . . . ?"), the interviewer's voice inflections, and by what the respondent thinks the interviewer wants to hear. Many people give answers they think are socially acceptable, whether the question has to do with income, religious beliefs, drug use, or education level.

In summary, telephone surveys are most appropriate when

- members of the population are very likely to have telephones;
- questions are relatively straightforward;
- experienced help is available; and
- quick turnaround is important.

Face-to-Face Surveys

Before the 1970s, face-to-face interviews were the *only* ones with any scientific credibility. They were thought to yield unrivaled response rates, allow for the lengthiest questionnaires, and collect the most accurate data.

Advances in mail and telephone surveys as well as high labor costs have taken the sheen off face-to-face interviews in the last fifteen years. However, surveys in which the respondent has one-on-one contact with the interviewer still offer enormous advantages under certain circumstances. Specifically, face-to-face interviews are uniquely suited to surveying populations for whom there is no list, or who are not likely to respond willingly or accurately by phone or mail.

A recent survey of workers in central Washington's food-processing industry illustrates how, sometimes, the only logical choice is to conduct face-to-face interviews. In this case, a community development organization wanted to evaluate how employment in local food-processing plants affected residents' income levels. Staff members from the organization knew of no address list that might be used to conduct a mail survey—neither vehicle registration nor utility lists were likely to be complete. In addition, they suspected that many people who worked at low-wage jobs in the processing plants did not have telephones. So, from the perspective of drawing a representative sample of residents, the researchers could conduct only a face-to-face survey.

Another reason in favor of conducting personal interviews was the education level of the people who were to be surveyed. Most workers in the processing plants used to be migrant farm workers who had only recently "settled out" of the migrant stream. Researchers working on the survey concluded that many residents would be unable or disinclined to complete a questionnaire that appeared in the mail and included questions about their household income.

Eventually, an "area frame" sampling technique was used to draw the sample. Researchers delineated a geographical area in which they were interested and, using maps from the Census Bureau, chose every tenth household for their sample. (Area frame sampling is described in detail in Chapter 5.)

Face-to-face surveys avoid the difficult problem of finding a complete list, but sometimes at a high cost in terms of money and time. Consider a worst-case example: The interviewer travels to a home in the area to which she has been assigned and arrives unannounced. (She can't call ahead to arrange a meeting because she has no phone number.) No one answers the doorbell, so she returns another day at a different time. This time, someone opens the door. The interviewer asks to speak to a particular person, say, the oldest member of the household. He is out of town and will not return for a week. The family has no phone, but now armed with a name and address the interviewer sends an introductory letter to the person she wants to interview. Upon returning to the house eight days later, the interviewer meets the respondent, who, after much cajoling, agrees to be interviewed a week later. Three visits and 16 days later, the interviewer finally fills out a questionnaire, at a very high cost to the project.

Expensive, time-consuming callbacks present the most serious problem when members of the sample are scattered over a large area. They are less of a problem in small communities. For example, a one-in-four sample from a town of 3000 households is much cheaper to conduct than a one-in-one hundred sample from a city of 75,000 households, even though both involve a sample size of 750.

Even more than telephone surveys, those conducted with the face-to-face method depend on interviewers who have been trained in why the research is being done, the format of the questionnaire, and sound interviewing techniques. It is possible to train inexperienced people in an intensive two- or three-day workshop before the survey begins. However, the supervisor's job will be much easier if he or she can find people who have worked as interviewers before, perhaps for the Census Bureau or other government agencies. The cost of using untrained people is a high degree of measurement error.

A good supervisor is also a *must* for face-to-face surveys. Even the best-trained interviewers run into problems that need immediate attention from a supervisor. And the work of those without experience must be monitored daily to make sure questionnaires are filled out completely and clearly. One interviewer who makes a consistent error over and over can ruin the accuracy of an entire survey. (How to supervise face-to-face surveys and train interviewers is described in Chapter 8.)

The temptation to cut costs in a face-to-face survey can be extremely high. Unfortunately, cost-cutting carries a high price in terms of error: Decreasing the sample size may inflate the sampling error; substituting someone besides the intended respondent (in order to avoid another visit) or using cheaper, unskilled interviewers will very likely increase measurement error. For these reasons, the face-to-face method should only be used when the project has an adequate budget.

Face-to-face surveys are expensive, but their strengths should not be overlooked. Interviewers have good control over who in the sampling unit serves as the respondent. They can increase the likelihood that people in the sample will agree to respond by explaining the importance of the survey and assuring them of its confidentiality. And they can make questions easier and less threatening by using visual aids such as flashcards that show answer choices.

All in all, face-to-face surveys are best suited to

- surveying populations for whom there is no list;
- collecting information from people who are not likely to respond willingly or accurately (or cannot be reached) by mail or telephone;
- complex questionnaires; and
- well-funded projects for which experienced interviewers and professional help are available.

Drop-off Surveys: A Convenient Hybrid

One final method to consider is the drop-off survey, in which people deliver questionnaires by hand to households or businesses. Respondents complete the questionnaires on their own and either return them by mail or leave them out to be collected.

To make the most of a drop-off survey, we recommend that surveyors leave questionnaires only with the intended respondents, rather than in mailboxes (which is illegal) or with people who must convey the purpose of the survey to someone else. Personal contact enables the surveyor to encourage respondents to complete the questionnaire. It also gives the survey a human face.

Drop-off surveys combine the low labor cost of mail surveys with the personal contact of face-to-face interviews. They are especially well-suited to

- small community or neighborhood surveys in which respondents are not spread over a large area;
- relatively short and simple questionnaires; and
- projects with a small staff but relatively large sample size.

What About Response Rates?

The term **response rate** refers to the proportion of people in a particular sample who participate in the survey. If 70 people in a sample of 100 people known to be eligible for the survey return a questionnaire, the response rate is 70 percent.

There was once a time when response rate was the main criterion used to choose among methods, and, almost always, face-to-face surveys won out. The mail and telephone options were usually rejected because researchers hadn't learned how to achieve high response rates without making personal contacts. Now, however, we understand better why people respond to surveys, and this understanding allows us to achieve equally high response rates using any of the three basic methods. For example, one can reasonably expect a 60 percent (or even higher) response rate in a mail survey of the general population, given the use of personalized cover letters, attractive questionnaires, and follow-up contacts. In well-organized surveys, similar rates can also be expected with the other methods.

Hence, response rate has become a much less important selection criterion. There is an exception, however, and it has to do with money. *The last few questionnaires that nudge the response rate to an acceptable level are the most expensive to secure.* Consider a face-to-face survey of 200 households. The first 100 interviews may be completed with a minimum of effort because the respondents are the easiest to contact. The next 100 interviews cost more because respondents aren't home the first few times the surveyor stops by, or they need more assurance of confidentiality from the project directors, or a host of other reasons. The cost of each additional interview increases quickly, so that the last 20 can cost more than the first 100! The project director may be tempted to stop trying after the first 100 interviews are complete—with a response rate of only 50 percent.

Compare that situation with a mail survey of the same 200 households. The cost of securing the last 100 responses is lower because follow-up work is done by mail, perhaps supplemented with phone contacts. The project director in this case is much less tempted to stop with a 50 percent response rate.

The key is this: With a fixed amount of money, a higher response rate is easier to achieve with telephone and mail surveys than with face-to-face interviews. When we disregard the cost issue, similar responses rates can usually be achieved with all three methods.

Remember the Budget

No matter how much we hammer on the idea of response rates and other error sources, the bottom line for most people is money. Therefore, you'll probably want to draw up several alternative budgets before deciding which survey method is best for you.

To help estimate how much your survey will cost, we have developed a hypothetical budget for each method. Without a doubt, *your costs will be different,* maybe one-third as much or maybe three times what we have estimated. There are many reasons why your numbers might differ from ours. You may, for example:

- need a smaller or larger sample size;
- have lower labor costs thanks to volunteers or paid staff who can be temporarily freed from other work;
- have a shorter or longer survey;
- have higher labor costs because you live in a high-wage area;
- be required to charge administrative overhead;
- have different printing costs or long-distance telephone charges;
- be able to use donated supplies; or
- need to spend money on additional contacts or incentives to raise your response rate (as discussed in Chapter 8).

Whatever the reasons, we encourage you to use our estimates only as grist for your own budget. Compare the assumptions we make with your own situation. Then look at each item and evaluate how your costs might vary.

Figures 4.1 through 4.3 show budgets for three hypothetical surveys, one for each of the basic methods. All three are estimated both with and without professional time. The numbers are based on assumptions about sample size, response rates, and labor costs. Be sure to compare our assumptions with your own situation. Note that only data collection and questionnaire-editing costs (discussed in Chapter 9) are included in these budgets. To estimate what the whole survey will cost, you need to add costs for data entry, analysis, and report writing.

For the mail survey, we assumed that a statewide sample is purchased from a survey research firm. The beginning sample size is 960. All 960 people are sent an advance-notice letter, a questionnaire with cover letter and stamped return envelope, and a follow-up postcard. Ninety percent of the second mail-out packets (about 865) are deliverable and the rest are returned, unopened. After the postcard follow-up, about 45 percent (or 390) of the deliverable questionnaires are completed and returned. That leaves 475 people who are sent a reminder letter with a new questionnaire and stamped return envelope. About 40 percent of these people (or 185) complete and return their questionnaires. The total return rate is 66 percent of deliverable questionnaires (575) less 10 percent unusables. Hence, the ending sample size in our hypothetical survey is roughly 520. Note that additional follow-up contacts or financial incentives intended to increase the response rate would add to the total cost. (See Chapter 8.)

As you can see, the mail survey budget looks very different from the other two. For mail surveys, we recommend a series of discrete tasks, each of which can be individually costed. It is much harder to make task-by-task estimates for telephone and face-to-face surveys. Hence, we roll the steps into one and calculate an average number of interviews per hour.

For the telephone survey, we assumed a beginning sample size of 1710. The sample is drawn from a telephone directory of several small communities using add-a-digit dialing, a technique explained in Chapter 5. Some in-state telephone charges are incurred because the sample includes people outside the local area market. Fifty-five percent of the numbers in the sample (about 940) belong to households that are eligible for the survey, and 55 percent of these are willing to respond. Hence, the ending sample size is about 520.

For the face-to-face survey, we assumed a beginning sample size of 985. The sample is drawn from maps of the relatively small community in which the survey is conducted. Seventy percent of the households selected (about 690) are eligible for the survey and can be

Figure 4.1

Estimated budget for a basic mail survey with an ending sample size of about 520 (see text for assumptions)

	Clerical hours @ $8.40/hr[a]	Other costs (dollars)	Total cost (dollars)	Your costs (dollars)
Prepare for survey				
Purchase sample list in machine readable form		375	375	______
Load data base of names and addresses	2		17	______
Graphic design for questionnaire cover (hire out)		100	100	______
Print questionnaires: 4 sheets, legal-size, folded, 1,350 @ $.15 each, (includes paper) (hire out)		203	203	______
Telephone		100	100	______
Supplies				
Mail-out envelopes, 2,310 @ $.05 each, with return address		116	116	______
Return envelopes, 1,350 @ $.05 each, pre-addressed but no return address		68	68	______
Letterhead for cover letters, 2,310 @ $.05 each		116	116	______
Miscellaneous		200	200	______
First mail-out (960)				
Print advance-notice letter	3		25	______
Address envelopes	3		25	______
Sign letters, stamp envelopes	6		50	______
Postage for mail-out, 960 @ $.29 each		278	278	______
Prepare mail-out packets	16		134	______
Second mail-out (960)				
Print cover letter	3		25	______
Address envelopes	3		25	
Postage for mail-out, 960 @ $.52 each		500	500	______

[a] $7.00 per hour plus 20% fringe benefits

Continued on next page

contacted. Seventy-five percent of these are willing to respond. Hence, the ending sample size is again about 520.

Costs for the three types of surveys vary considerably. Excluding professional time, the estimate for a mail survey is $4883; for a telephone survey, $6919; and, for a face-to-face survey, $16,570. On a per-completed-questionnaire basis, the estimated costs are $9.39, $13.30, and $31.87, respectively.

Figure 4.1
Continued

	Clerical hours @ $8.40/hr[a]	Other costs (dollars)	Total cost (dollars)	Your costs (dollars)
Postage for return envelopes, 960 @ $.52 each		500	500	______
Sign letters, stamp envelopes	12		100	______
Prepare mail-out packets	14		118	______
Third mail-out (960)				
Prestamped postcards, 4 bunches of 250 @ $.19 each		190	190	______
Address postcards	3		25	______
Print message and sign postcards	6		50	______
Process, precode, edit 390 returned questionnaires, 10 min each	65		546	______
Fourth mail-out (475)				
Print cover letter	3		25	______
Address envelopes	3		25	______
Sign letters, stamp envelopes	3		25	______
Prepare mail-out packets	20		168	______
Postage for mail-out, 475 @ $.52 each		247	247	______
Postage for return envelopes, 475 @ $.52 each		247	247	______
Process, precode, edit 185 returned questionnaires, 10 min each	31		260	______
Total, excluding professional time	196	3240	4883	______
Professional time (120 hrs @ $35,000 annual salary plus 20% fringe benefits)		2423	2423	______
Total, including professional time		5663	7306	______

[a] $7.00 per hour plus 20% fringe benefits

For several reasons, these estimates are far lower than what is likely to be charged by organizations who contract to do data collection. First, we have assumed that our readers have available or can obtain the necessary office space and equipment to do a survey. A professional survey organization would have to charge overhead to cover such costs. Second, it would have to cover costs of negotiating and communicating with the client, as well as salaries and wages for all

Figure 4.2

Estimated budget for a basic telephone survey with an ending sample size of about 520 (see text for assumptions)

	Clerical hours @ $8.40/hr[a]	Interviewer hours @ $6.48/hr[b]	Other costs (dollars)	Total costs (dollars)	Your costs (dollars)
Prepare for survey					
Use add-a-digit calling based on systematic, random sampling from directory	10			84	______
Print interviewer manuals	2		20	37	______
Print questionnaires (940)	4		50	84	______
Train interviewers (12-hour training session)		108		700	______
Miscellaneous supplies			25	25	______
Conduct the survey					
Contact and interview respondents; edit questionnaires; 50 minutes per completed questionnaire		430		2786	______
Telephone charges			3203	3203	______
Total, excluding professional time	16	538	3298	6919	______
Professional time (120 hrs @ $35,000 annual salary plus 20% fringe benefits)			2423	2423	______
Total, including professional time			5721	9342	______

[a] $7.00 per hour plus 20% fringe benefits.
[b] $6.00 per hour plus 8% F.I.C.A.

employees involved in the project. Even without including charges for coding, data entry, analysis, and report writing, the contractors' costs could easily be twice as high as our estimates.

Remember, too, that the face-to-face survey on which we based the third budget is a local project that does not entail complicated sampling or high travel and interviewer costs. This is a very different design than those conducted by most professional survey organizations.

Paradoxically though, the estimated costs in our budgets are much higher than those that have actually been incurred for similar surveys

Figure 4.3

Estimated budget for a basic face-to-face survey with an ending sample size of about 520 (see text for assumptions)

	Clerical hours @ $8.40/hr[a]	Interviewer hours @ $8.10/hr[b]	Other costs (dollars)	Total costs (dollars)	Your costs (dollars)
Prepare for survey					
Purchase map for area frame			200	200	______
Print interviewer manuals	2		12	29	______
Print questionnaires (690)	4		345	379	______
Train interviewers (20-hour training session)		140		1,134	______
Miscellaneous supplies			25	25	______
Conduct the survey					
Locate residences; contact respondents; conduct interviews; field edit questionnaires; 3.5 completed interviews per 8-hour day		1192		9,655	______
Travel cost ($8.50 per completed interview; interviewers use own car)			4420	4,420	______
Office edit and general clerical (6 completed questionnaires per hour)	87			728	______
Total, excluding professional time		931332	5002	16,570	______
Professional time (160 hrs @ $35,000 annual salary plus 20% fringe benefits)			3231	3,231	______
Total, including professional time			8233	19,801	______

[a] $7.00 per hour plus 20% fringe benefits.
[b] $7.50 per hour plus 8% F.I.C.A.

done by local organizations. Volunteers were used to do the work, and other resources were donated. Once again, we encourage you to consider these budgets as only starting points for calculating the costs of doing your own survey.

What Does the Future Hold?

There are times when no single method seems just right for a particular survey. For example, survey organizers might not have access to a complete population list for a mail questionnaire, and they might be unable to afford the next best alternative, a face-to-face survey. Or they might need a response rate of 80 percent, which they cannot expect to achieve with any one method alone.

Increasingly, professional survey researchers have turned to **mixed mode** surveys to solve problems like these. *Mixed mode* means using two or more methods of data collection for a single survey. Typically, one starts with the method that best solves the problem of adequate coverage and/or cost, and then switches to a second or third method to get the highest possible response rate.

For example, the U.S. Census Bureau recently conducted a national survey of scientists and engineers (Shettle 1993). Researchers obtained a sample of names from Census records and contacted people on the list four times by mail. (The first letter included a $5 check.) Next, they tried to telephone people who had not responded, in some cases because addresses on the list were outdated. Finally, they attempted to conduct face-to-face interviews with people who couldn't be reached by telephone. Sixty-two percent of the sample responded after the four mailings, an additional 12 percent responded when contacted by telephone, and 8 percent more responded when asked to give a face-to-face interview. The total response rate was 82 percent.

Another example of a mixed mode survey is the decennial census, for which the Census Bureau sends a mail questionnaire to every household on its master population list. People who do not respond are then contacted by face-to-face interviewers, and some may be telephoned.

The usual strategy in a mixed mode survey is to get the highest response rate possible with the least expensive method first and then switch to more costly methods. Since some people who refuse to participate with one method are likely to respond to another, mixed mode surveys almost always result in higher response rates than single mode surveys.

A word of caution: A step-by-step explanation of how to conduct mixed mode surveys is beyond the scope of this book. Among the many problems that can arise is the method effect whereby people answer certain questions one way on self-administered questionnaires and another way in an interview. Hence, we recommend that you get professional help if you think a mixed mode survey is the only way to get the needed information.

Mixed mode surveying is not the only recent advance in data collection. Some researchers are experimenting with questionnaires on computer diskettes. Others have used fax machines to distribute and collect survey forms. Still others have asked respondents to enter their answers to prerecorded questions on Touch-Tone telephones. As of yet, we doubt that any of these methods will enable most of you to get adequate response rates in your surveys. However, twenty years ago most naysayers said, "You can't do surveys by telephone!" Just as their pessimism proved unwarranted, so the new methods mentioned here may gain widespread use for certain types of surveys in coming years.

For more detail on choosing a survey method, see the following.

Chapter Two, "Which Method is Best: The Advantages and Disadvantages of Mail, Telephone, and Face-to-Face Surveys," by Don A. Dillman, *Mail and Telephone Surveys: The Total Design Method*, Wiley-Interscience, New York, 1978.

5

When and How to Select a Sample

A **sample,** as we use the term here, is a set of respondents selected from a larger population for the purpose of a survey. The main reason to sample is to save time and money. Drawing a sample is sometimes quite simple, as in the case of a recent survey conducted by a school superintendent. She wanted to learn why voters had turned down a bond levy just one week before. She knew that the **population** she wanted to survey included all registered voters in the school district. The county clerk's office gave her a complete voter registration list containing 5212 names and addresses. She then randomly picked a number between 1 and 10, which was 6. Starting with this random choice—the sixth entry on the list—her clerical assistant selected a **systematic sample** by entering every tenth name into a computer.

Because the bond levy was a high-profile issue of importance in the community, the school superintendent expected to get responses from about 75 percent, or 390, of the 521 registered voters in the sample. With this many completed questionnaires, the expected sampling error would be plus or minus (±) 5 percent, which she had decided would be acceptable for her particular information needs.

In this straightforward example, sampling involves deciding how much sampling error is acceptable and therefore, how large the sample needs to be. Three specific steps are involved: defining the survey population, obtaining an adequate population list, and then selecting the sample. The superintendent did each of these three things.

Unfortunately, sampling is not always this simple. Sometimes, it isn't even necessary, a situation we address first in this chapter. Other times, circumstances require sampling methods far more complex than we can discuss here. Hence, our goal here is to give perspective on sampling issues so that you can either decide to use the relatively simple methods described in this chapter or seek additional help.

When Is Sampling Useful?

As we emphasize in this book, the power of sample surveys is their ability to obtain information from a relatively few respondents to describe the characteristics of an entire population. What we gain by sampling is efficiency: It takes less time and money to interview a few respondents than to interview many.

However, sampling is not always necessary. When the study population is very small, efficiency may not be a big concern. For example, in a mail survey of the 250 PTA members at an elementary school, sampling might offer few cost advantages. Perhaps we are interested in how this small group of people feels about a new school policy. If we want to estimate how many support the policy and how many do not, and if we want our estimate to have a sampling error of only 3 percent in either direction, we need about 200 completed questionnaires. (This supposes PTA members are about evenly split on the issue, a "conservative" assumption we'll explain later.) Since the cost of sending out and retrieving 50 additional questionnaires is probably not high, we might as well forget sampling and send a survey to all 250 people.

Still, there are several reasons why sampling might be appropriate even for a small group like the 250-member PTA. First, if we are doing telephone or face-to-face interviews instead of a mail survey, 50 fewer interviews would mean a substantial savings of time and money. Second, regardless of the survey method, if we can tolerate a higher sampling error, we can get by with a much smaller sample and therefore a cheaper survey. For the 250 PTA members, completed questionnaires from a random sample of only 70 respondents would allow us to make estimates with a sampling error of about ±10 percent. Sampling in this case would make good sense. Third, if we expect the population to have less variation (in other words, if most people have the same opinion and just a few have another), our sample can be even smaller.

So the question of whether or not to sample depends on survey method, population size and variation, and our need for precision. These issues bring us directly to the question of sample size.

How Large Should a Sample Be?

Luckily, choosing a sample size does not have to be done in an arbitrary way. All it takes is organizing what is known about the study population and the survey objectives. Specifically, sample size depends on:

- how much sampling error can be tolerated;
- population size, *if* the population is small (and how "small" depends on how much precision is required);
- how varied the population is with respect to the characteristics of interest; and
- the smallest subgroup within the sample for which estimates are needed.

Table 5.1 gives the sample sizes necessary to estimate population percentages, given various levels of sampling error, population size, and variation. For the PTA example discussed above, we read the table as follows: Obtaining 203 completed questionnaires allows us to be

Table 5.1

Final sample sizes needed for various population sizes and characteristics, at three levels of precision

	Sample size for the 95 percent confidence level					
	±3% sampling error		±5% sampling error		±10% sampling error	
Population size	50/50 split	80/20 split	50/50 split	80/20 split	50/50 split	80/20 split
100	92	87	80	71	49	38
250	203	183	152	124	70	49
500	341	289	217	165	81	55
750	441	358	254	185	85	57
1,000	516	406	278	198	88	58
2,500	748	537	333	224	93	60
5,000	880	601	357	234	94	61
10,000	964	639	370	240	95	61
25,000	1,023	665	378	234	96	61
50,000	1,045	674	381	245	96	61
100,000	1,056	678	383	245	96	61
1,000,000	1,066	682	384	246	96	61
100,000,000	1,067	683	384	246	96	61

How to read this table: For a population with 250 members whom we expect to be about evenly split on the characteristic in which we are interested, we need a sample of 152 to make estimates with a sampling error of no more than ±5 percent, at the 95 percent confidence level. A "50/50 split" means the population is relatively varied. An "80/20 split" means it is less varied; most people have a certain characteristic, a few do not. Unless we know the split ahead of time, it is best to be conservative and use 50/50.

Numbers in the table refer to completed, usable questionnaires needed for various levels of sampling error. Starting sample size should allow for ineligibles and nonrespondents. Note that when the population is small, little is gained by sampling, especially if the need for precision is great.

Remember! Sampling is just one of the four sources of error and is the only one whose effect we can usually estimate with confidence (see Chapter 2).

95 percent confident that our estimates will have a sampling error no more than ±3 three percent (as long as we think the PTA members are split about evenly on the characteristic we're interested in). If we can tolerate more error or less confidence, or if we think the PTA members are less varied in their opinions (say 80/20 instead of 50/50), we can get by with a smaller sample.

A more careful, statistical description of what we mean is this: 19 out of 20 times—or 95 percent of the time—that we have a random sample of 203 from our population of 250, a range that is the sample estimate ±3 percent can be expected to contain the population value for all 250 people.

For example, let's say that 55 percent of the sample favors an increase in teachers' salaries. "Fifty-five percent" is our *estimate* of the proportion of the entire PTA population that favors salary increases. The chances are 19 out of 20 that the population value is within 3 percent of the estimate, in either direction—in other words, between 52 percent and 58 percent.

Table 5.1 illustrates something remarkable about sampling. For large populations, we need samples of about the same size to make our estimates, regardless of whether we are interested in 100,000 people or 1,000,000. Only when we are concerned with groups of less than several thousand does the sample size make a big difference. That explains why national and state polls cited by news organizations often base their results on 1100 to 1200 interviews of people. That is roughly the sample size needed from a large population to make estimates of percentages with a sampling error of no more than ±3 percent at the 95 percent confidence level.

In our PTA example, we noted how our expectations about diversity of opinion affected sample size. To understand why, consider this: To find out what blood type you have, a nurse theoretically needs to sample only *one* of the millions of blood cells in your body—because all your cells are the same type. In other words, they are completely uniform. Contrast blood cells in a single body with fingerprints in a population of 1000. In this case, a sample of one person would tell us nothing about fingerprint patterns in the population. Only a *complete census* would tell us what the patterns are.

The survey research problems we deal with in this book fall between such problems as illustrated with the blood- and fingerprint-typing examples. The reason is that populations on which social and economic studies focus are rarely completely uniform or totally distinct. The key is that more diverse or variable populations require larger sample sizes and vice versa. If researchers have no prior knowledge about how diverse their target population is, we recommend taking a conservative approach with respect to sample size. That means that they should assume more diversity on dichotomous (two-sided) characteristics of interest to the study (in other words, a 50/50 split). Appendix 5.A explains in detail the relationship between variation, sample size, and sampling error.

We need to make several very important qualifications to what we've said so far. First, researchers are almost always interested in subgroups within the population they are studying. Gallup wants to know about voters in particular states, the Bureau of Labor Statistics needs to estimate unemployment rates for various age and racial groups, and the PTA needs information about parents of children in different grade levels. These researchers should base their decision about sample size not on how big the entire population is but on how big their subgroups are. For example, if Gallup surveyors want to compare voter preferences in Florida and Oklahoma, they need an adequate sample size from each state. If the Bureau of Labor Statistics wants to estimate unemployment among young black males, it needs a big enough sample from this particular subgroup of the population, and so on.

Another way to look at this is to say that sampling error depends on the size of the subgroup. From Table 5.1, we know that if we have a completed sample of 516 from a population of 1000 people, our sampling error is ±3 percent on the whole sample (assuming a roughly even split), but ±5 percent when we look at just half the sample at a time. In other words, we give up accuracy when we start breaking our sample into separate parts for analysis. If we are unwilling to sacrifice accuracy, we must increase sample size.

The second important qualification we need to make has to do with the actual sample size on which sampling error is based. Table 5.1 gives us the number of *completed, usable interviews* necessary for the various levels of sampling error. In other words, we figure our sampling error on how big the sample ends up being after we take out people who are ineligible or refuse to participate in the survey and after we discard illegible questionnaires. We never expect to end up with as many interviews as we have in our original sample. The safest way to make sure we end up with enough usable cases is to figure out the number of questionnaires needed in the final sample, and then work backward. For a mail survey in which a telephone directory is used as a frame, we might want to end up with 601 usable questionnaires. Assuming that 90 percent of the addresses are usable, that 70 percent of the remaining households respond, and that 10 percent of the returned questionnaires are illegible or incomplete, we need a starting sample of about 1060 ($601 \div 0.9 \div 0.7 \div 0.9 = 1060$).

In a roundabout way, we have come back to the question of whether or not to sample. Consider the example of an economic development agency doing a mail survey of their state's small businesses, of which there are 2500. The agency's staff people want to know how many of these businesses market out-of-state. They think the proportion is no more than 20 percent but they need facts to back up their hunch.

Should the researchers sample from the population or survey everyone? From Table 5.1 we know that, if they decide to sample, they need 224 usable questionnaires to get a sampling error of ±5 percent, assuming an 80/20 split on in- and out-of-state marketing. A 10 percent, or 1-in-10, sample would give them a list of 250 businesses. If they can get a 90 percent response rate, which is not out of the question if they do intensive follow-ups by phone and mail, they'll have 224 questionnaires with little indication of a potential nonresponse problem.

If they send the survey to all 2500 businesses instead of sampling and *don't* follow up to raise the response rate above, say, 9 percent, they could still end up with only 224 questionnaires. Their sampling error would still be ±5 percent, but now they would have a potentially serious problem with nonresponse error, a problem they cannot properly evaluate because they know nothing about how similar their respondents are to those who were asked but didn't respond. In other words, they can get far more accurate estimates if they do a good job with a sample than if they do a poor job with a census. If they are interested in accuracy, a 90 percent response rate from a sample would be much better than a 10 percent response rate from a survey of everyone.

The Three Steps in Sampling

If you decide that sampling is appropriate for your survey, you have to do three things: identify the target population, put together a population list, and select the sample.

Step one is to identify the target population as precisely as possible and in a way that makes sense in terms of the purpose of the study. It is important to be specific enough that everyone involved in the research knows who is eligible for the survey and who is not. Anticipating all the unusual cases is impossible, but the more precise the definition is, the easier it will be to handle questions once the survey begins.

The second step in sampling is to find or put together a list of the target population. This is the list from which the sample will eventually be drawn. Survey statisticians call it the **list frame**. There are many kinds of lists: telephone directories; club membership lists; customer lists from utility companies; public agency client lists; voter registration lists; and so on.

If you can neither find an existing list nor construct one yourself, you need an alternative. One possibility is to use an **area probability sampling frame** approach, in which small geographical areas are sampled from within the frame. The entire geographic area covered by the frame might be a census tract, township, or even a whole country.

The third step is to actually select the sample. Sampling methods range from simple to extremely complex; the latter are beyond the scope of this book. But for many surveys of small populations and small areas, uncomplicated designs like simple random sampling and systematic sampling are adequate (as described later in this chapter). Complex methods become necessary when large areas are being surveyed and efficiency is paramount. They are also necessary when researchers want to make precise estimates about relatively small subgroups within large populations.

We look more closely at the second and third steps in the next two sections of this chapter.

Finding Good Lists

Finding a list from which to draw a sample is sometimes easy, especially for surveys of small, specific populations. For example, you might get a list of parents of local school children from school administrators, a list of doctors who practice medicine in your state from a licensing board, or a list of registered voters from the county elections office.

General population lists pose a greater challenge, especially at the national and state levels, where they are simply not available. (The Census Bureau compiles an address list for the decennial census but Title 13 of the U.S. Code prevents making it public.) At the local level, you can consider using telephone or city directories, lists of utility hookups, or voter registrations lists.

An important point that can't be overemphasized is this: A critical part of getting lists is having official sponsorship. In Chapter 3, we stressed the importance of drawing into the planning process the people who will use or be affected by the study results. If these people are part of an agency or organization with a direct interest in your work, they may be willing to cosponsor the project and help you get a list. Alternatively, you may want to involve university or college researchers in your project. They too may give the credibility you need to obtain lists that would otherwise be unavailable.

Even as university researchers, we have been told many times that we cannot obtain the list needed for a particular survey. And just as many times, we have found someone who *can* get it—a local mayor, a school board chairperson, or an agency administrator. All it takes is getting one of these key people to recognize the value of the project and then working through official regulations.

Evaluating whether we have found a "good" list is another issue. As we'll see later in the chapter, random samples give every member of a population an equal and independent chance of selection. (That's what makes them probability samples.) Therefore, the very best list is one in which every member of the target population is listed once and only once. Such a list allows us to select a random sample without worrying whether we over- or undersample any particular group. In other words, coverage error is not a concern if the list is a good one.

Unfortunately, few lists are perfect. The majority omit certain members of the target population, include them more than once, or include people not of interest to the study. The result is that the people who are the focus of the survey are not the same as those who are actually on the list.

Two likely lists for local, general-population surveys illustrate how coverage error might be a problem. The first is the telephone directory. People without phones and those with unlisted numbers or new listings have no chance at all of being selected. In addition, households with two or more phone numbers, or in which spouses with different last names are both listed, have a greater chance. Hence, people have *unequal* chances of being selected from a telephone directory.

The second candidate is the utility hookup list, which poses similar problems. For example, if rental property is listed by landlord, tenants have no chance of being selected in the sample. Furthermore, landlords who own more than one property and who may not even live in the community have a greater chance of selection than single-property home owners. In short, random selection from such a list will not yield a representative sample unless the list is corrected.

General-population lists are not the only ones vulnerable to coverage error. Special population lists (for example, membership lists for professional associations and directories for nonprofit groups) are often out-of-date, incomplete, or simply contain wrong information. Even voter registration lists, which one might think of as being relatively error-free, often include inaccurate addresses and do not reflect people who have recently moved in or out of the area.

Omissions, duplicate entries, and inaccuracies are not always a problem, *but every list must be considered individually.* An acceptable list for one survey may be out of the question for another. For example, phone directories typically pose fewer problems in rural than in urban areas because rural people are not very likely to have unlisted numbers. Aside from the problem of people without phones—who may or may not be important in the context of a particular study—the directory may be acceptable for a telephone survey of residents in a small community.

When Finding a List Seems Impossible

People often want to survey populations that simply are not listed anywhere. Here are some examples:

- **Visitors at a museum or shoppers at a grocery store.**

Populations like these consist of people who show up at a particular location. There may be more than one entrance or exit and more people at certain times than others. The best strategy is to sample people as they arrive or leave. One way of doing it is to sample hours of the day (or week). Then during the sampled times, sample every *n*th person (for example, every 10th or 15th). This method requires having enough interviewers so that the same proportion of people are interviewed during busy and slow times.

More efficient sampling methods may be available, but we suggest you consult with someone—for example, a sampling statistician—who understands *time-and-location* sampling to see what might work best for you.

- **Households with someone over 55 years of age or individuals who have traveled outside the state in the last year.**

In cases like these, there is no way to know in advance whether a household or individual will be eligible for the survey. The usual solution is to ask a few questions at the beginning to **screen** respondents for eligibility. Whether screening is workable depends on how many contacts will be required to achieve the desired sample size. For example, locating people who are legally blind by phoning a general-population sample would take an enormous number of calls.

Screening works best in telephone and face-to-face surveys. We recommend against using it in mail surveys since people who are ineligible usually don't return their questionnaires. That makes it impossible to differentiate between refusals and ineligibles and, therefore, to calculate the response rate.

- **Minority-owned restaurants or firms that provide package delivery service.**

There are no lists for populations like these, and nonlist sampling techniques may be too complicated and expensive. This problem calls for creative solutions, which usually means developing one's own list from multiple sources. Sometimes the Yellow Pages is a place to start. Other times it works to contract organizations whose members are likely to include some of the target population.

Coverage error can be a serious problem whenever a list is compiled from several sources. For this reason, you must ensure that the list is as complete as possible. Ask people who are knowledgeable of the target population if they can identify people who should appear on your list. And if another organization has what seems to be a good list but cannot make it available, ask them to check your list against theirs to identify coverage error. Additionally, if your list includes people who are ineligible for the survey, encourage them to return the questionnaire so that the response rate can be calculated accurately.

Uncomplicated Sample Designs

Once you have a list, you're ready to sample. The most basic method, **simple random sampling** (SRS), gives each member of the target population an equal chance of being selected. It requires that all members of the target population be included on a list. There are three ways to select a simple random sample: in a lottery, in other words, by picking out of a hat; with a random numbers table; and with a computer-generated list of random numbers (Henry 1990).

Here is an example of a survey that involved SRS from a list:

> A local school board surveyed parents about the need for improving teacher-parent communication. Their target population consisted of all parents of children enrolled in grades K–12. The principal at each of the district's six schools provided the board with a list of all parents of children enrolled at his or her school. The list was current as of the beginning of the school year and included both names and addresses. A careful check was made for duplicate listings (parents of children in more than one school). The board then used a random numbers table to select a 20 percent sample of families and mailed one questionnaire to the parent or set of parents in each sampled family.

The school board used a random numbers table from a statistics textbook. They could also have used one of many statistical software packages or pocket calculators that generate random numbers.

Unfortunately, SRS is cumbersome when combined with a long list. A good, simple alternative is to use systematic sampling with a random start. Like SRS, systematic sampling gives all members of the target population an equal chance of being selected. However, in this case, only the first element in the sample is chosen from a random numbers table. After that, elements are chosen systematically. For example, in a one-in-ten sample, the first element is selected randomly from elements 1 through 10, and every tenth element is chosen from then on. In a one-in-five sample, every fifth element is chosen, and so on.

Probability and Nonprobability Samples

Random and systematic sampling are examples of **probability** designs. In the parent survey described above, the school board drew a probability sample. Its distinguishing feature was that every parent had a known chance of being selected by virtue of being on the list. To maintain the integrity of the selection process, it was very important that the school board follow a well-defined selection process in which subjective judgment played no role. They could not drop a parent out of the sample because they expected that he or she would not respond or was known to be unpleasant. Had they done so, they would have had a **nonprobability** or **purposive** sample.

Nonprobability sampling depends on subjective judgment. The surveyor selects a sample because it is convenient, because he or she believes it is "typical," or perhaps because it is composed of especially interesting cases. Some members of the population have a very high

How to Use a Random Numbers Table

To use a random numbers table with a list frame, start by numbering every entry on your list from 1 to the end of the list. In the school board example, the population list included 1878 parents or sets of parents, so the numbered list looked like this:

0001	Mary Johnson
0002	Beth and Sam Smith
0003	Jan and Mark Lee
⋮	
0900	Suzanne Smarthton
0901	Ann and Joe Felman
⋮	
1877	John Carpenter
1878	Tom Waller and Sarah Steele

Next, randomly choose a starting place in a random numbers table. (See, for example, pages 622–623 in Kish's *Survey Sampling,* cited at the end of this chapter.) It doesn't matter where you start because the table is entirely random.

Now proceed through the table and select each entry on your list whose number matches. You'll only need to look at as many digits in the random numbers table as you have in your highest ID number. The school board used the last four digits from each random number because their highest ID number was 1878. Skip all random numbers that don't have a match in your list. Stop when you get the number of entries you need to complete your sample.

Here's what happened when the school board matched numbers from the table to their population list:

Random No.	**Parent ID**	**Parent Name**	
800900	0900	Suzanne Smarthton	1st
421998	no match		
932183	no match		
090001	0001	Mary Johnson	2nd
451877	1877	John Carpenter	3rd
142091	no match		
.			
.			
.			
220901	0901	Ann and Joe Felman	376th

The school board wanted a 20 percent sample, so they repeated the matching process until they had 376 (one-fifth of 1878) parents or sets of parents.

(Source: Adapted from Henry 1990.)

chance of selection, others have no chance at all, and we have no way of positively knowing the probability in either case.

An example of a nonprobability sample is the way focus groups are usually chosen. Members of focus groups are carefully (but not systematically) selected because of some characteristic they have. Selection is based on judgment rather than according to a systematic method that would give everyone in the study population a known chance of being chosen.

Nonprobability or purposive sampling is appropriate in certain circumstances, especially for exploratory research intended to generate new ideas that will be systematically tested later. It may also be appropriate for surveys conducted to organize communities, identify leaders, or build networks. However, it is imperative that people avoid using judgmental or nonprobability samples in survey research if the goal is to learn about a larger population. In contrast to a probability sample, we have no way of knowing the accuracy of a nonprobability sample estimate. It might be accurate, but then again, it might not. Hence, whatever new information is gained through the research applies only to the sample itself. Nonprobability sampling simply can't be defended to would-be users of the results.

Selecting a Respondent from within a Household or Business

Sometimes, drawing a sample involves only the process we have discussed so far—selecting individuals, households, or businesses either from a list or through some other random process. This is often the case in surveys of special populations for which a list is available. In a survey of trade association members, you might have a list of business owners complete with names, addresses, and phone numbers. In that case, sampling is a one-step procedure. You simply select respondents from the list and ask for them by name when you call, stop by, or write. In a telephone interview, you might say something like:

Hello. Is this Jeff Carpenter?

[If YES] Are you still the owner of Southeast Construction?

If the answer is YES again, you have identified the respondent and can proceed with the interview. However, especially in general-public surveys of all adults, a second step is necessary to select the respondent because one rarely wants to interview just anyone, regardless of age, gender, or other characteristics.

Sometimes any *adult* will do, because he or she can speak for the household and doesn't need to be chosen randomly or screened for a particular characteristic. In this case, the respondent can often be anyone over 18 and no further selection process is needed. By interviewing a household spokesperson, you end up with a sample of all households, not a sample of all adults. This is fine for a survey of household recycling practices (for example) but not fine for a survey of individual voting preferences.

Sometimes, surveyors are interested in a respondent who has a unique characteristic within the household or business, for example, the oldest member of the household or the person who does the bookkeeping in a business. To identify a particular person, the questionnaire should begin with a screening question. For example, in a telephone survey about child care, interviewers might introduce the survey and then continue as follows:

> I need to speak to the person in your home who can tell us about your child care arrangements. Would that be you or someone else?

And finally, there are times when surveyors need a probability sample of all adults, in other words, each adult member of sampled households must have an equal chance of being selected. In this case, we recommend asking for the person who is 18 years of age or older who had *the most recent birthday.* This is the simplest and least threatening way of randomly selecting from within a household. After a proper introduction, an interviewer might say:

> We need to be sure we give every adult a chance to be interviewed for this study. Thinking only of adults in your household—that is, people 18 years of age or older—which one had the most recent birthday?

Another procedure is the Kish method. It involves identifying the number, gender, and age of household members. It can be difficult to use in mail surveys, and in telephone surveys some researchers believe it can produce higher telephone refusal rates than other methods.

How the Survey Method Affects Sampling Frame and Design

We have spent much of this chapter discussing lists: where to find lists, how to evaluate them, and how to use them to draw samples.

Mail surveys cannot be done without lists, and *successful* mail surveys cannot be done without *good* lists. As we said earlier, coverage error can be a major problem if mailing lists have a significant number of omissions, duplicate entries, or inaccuracies.

If you use any other survey method, however, you can work around incomplete lists or use an entirely different kind of frame. We turn now to how these alternatives pertain to telephone and face-to-face surveys.

Special Considerations for Telephone Surveys

For phone surveys of special populations, sampling from an available list is a realistic option. For example, when the Oregon Department of Fish and Wildlife wanted to study whether people were fishing in certain parts of the state, the agency used phone numbers from a list of people who had recently purchased fishing licenses. When researchers at Washington State University wanted to know whether faculty, staff, and students were satisfied with services provided at the student union building, they compiled a phone list from the student directory. Although they almost always need cleaning up to correct inaccuracies and weed out duplicate entries, lists for surveys of special populations like these are usually available or can be constructed.

General-public telephone surveys pose more serious problems. The list most likely to be considered is the phone book. In smaller communities, it may be practical to randomly sample from a telephone directory or, to save time, to draw a systematic sample. However, because people with unlisted phone numbers or new listings have *no chance of being selected* from a directory, coverage error can be a serious concern.

Random digit dialing (RDD) is a technique commonly used to get around problems of unlisted phone numbers and out-of-date directories. Briefly, this is how RDD works in a small community: The surveyor looks in the phone book to find out what 3-digit prefixes, or "central office codes," are used in the study area. Assuming he or she doesn't want to exclude any of these codes from the sample, the next step is to generate (preferably with a computer) 4-digit random numbers to use with the 3-digit codes. All of the 7-digit numbers on the resulting list are then called for interviews. Nonworking and ineligible numbers are eliminated from the sample as they are encountered (Frey 1989).

In many areas, only a few blocks of 1000 telephone numbers are used for the general public. Other blocks are not used at all or are assigned to particular establishments, such as a university. RDD is most efficient when surveyors know ahead of time which blocks contain working numbers. For example, in Colfax, Washington, only three blocks of 1000 are assigned—the 2000, 3000, and 4000 blocks. Since only one 3-digit code—397—is used in Colfax, all possible working numbers are between 397-2000 and 397- 4999. Therefore, if surveyors were using RDD to generate a sample for their telephone survey of Colfax, they would use only random numbers between 2000 and 4999. (Even within the range, less than one-third of the numbers are working.)

For general-public surveys in larger areas, researchers can purchase lists of telephone numbers that have been randomly generated by private firms. Many companies advertise that their lists have been developed to include as few nonworking numbers as possible.

In addition to random digit dialing, there is a second technique that can be used to avoid coverage error problems associated with unlisted numbers and out-of-date directories. It is called **add-a-digit sampling**. This technique involves drawing a random or systematic sample from the directory and then adding a randomly chosen number from 1 to 9 to the last digit of each number in the sample. One way to accomplish this is to add the same digit to each phone number. Hence, if 883-0527 is selected and 5 is the random digit, 883-0532 is called. Another method is to generate separate random digits for each number sampled from the directory.

The advantage of add-a-digit sampling is that it provides a way of including new listings and unlisted phone numbers in the sample. It yields fewer nonworking numbers and is therefore more efficient when the phone company assigns numbers in banks and leaves large blocks unassigned (Frey 1989). This is especially true in small communities with so few households that most of the possible numbers are unassigned.

For local surveys in communities where unlisted numbers and out-of-date directories are a problem, we recommend using add-a-digit sampling. It is relatively efficient and easy to understand. Some evidence suggests there *may* be a problem with this method; that is, one can't be sure that an add-a-digit sample adequately represents all unlisted numbers and new listings in the population (Lepkowski 1988). For telephone surveys where great confidence in accuracy is required, add-a-digit sampling is not recommended.

Our suggestion: Use add-a-digit sampling with caution. Talk to someone from the phone company about whether new listings are in blocks of numbers that don't appear in the most recent directory. Since we don't expect our readers to be doing large surveys in which great precision is essential, the results obtained from add-a-digit dialing should be satisfactory, and the gains in simplicity, significant.

For those readers interested in other techniques and for those doing telephone surveys of larger areas, we strongly suggest consulting one of the several excellent reference books that are listed at the end of this chapter.

Special Considerations for Face-to-Face Surveys

If you ask a survey statistician for advice on sampling for a face-to-face survey, he or she may say you shouldn't even attempt it. Indeed, drawing national, statewide, or even some substate samples for face-to-face interviews is very complicated. Such samples are usually drawn in several stages and involve a process called *clustering*—all designed to avoid interviewer travel costs that would be involved in a random sample spread over a large geographic area. These techniques are beyond the scope of our discussion here.

However, it *is* practical for you to consider using a simple random sample for face-to-face interviews if you will be working in a small area, for example, in a single neighborhood or community. Then, the cost of interviewing a random (or systematic) sample will likely be the same as for a multistage, cluster design. Small-area sampling for face-to-face surveys is easiest and least expensive if a complete list of addresses can be obtained.

As in the case of phone surveys, lists are easiest to get for special populations, for example, all banks or teachers in a community. Even for general-population surveys, one can often get a utility hookup list if someone with authority at the utility company believes the survey is worthwhile. If the study area is small enough, it is also possible to create a list by walking up and down each street and recording addresses, taking special care to be sure that apartments in multiunit buildings are correctly identified.

If the area is relatively small but no list is available, sampling is more complicated. Earlier in this chapter, we introduced the idea that samples can be drawn not only from lists, but also from geographical areas. Map-based "area frames," as they are called, are commonly used when no list is available. Once in a while, it is possible to find community or neighborhood maps that depict residential and commercial buildings. These maps can be used as area frames from which household samples can be drawn in either a random or systematic way.

Since maps depicting buildings are rare, researchers who do household surveys often use county block maps produced by the U.S. Bureau of the Census. The Bureau divided the entire nation into about seven million blocks for purposes of conducting the 1990 decennial census. In urban areas, census blocks correspond to city blocks and have an average population of about 85 people. In rural areas, block boundaries are visible features such as roads, power lines, and fences. Their average population is about thirty. Block maps can be ordered for individual counties from the regional census offices listed in Appendix 5.B.

To sample blocks in your study area, first find out how many occupied housing units are in each block. (This information is available from the State Data Center; call your regional census office for the phone number.) Then figure out how many households there are area-wide, and what fraction needs to be sampled to get the desired sample size. What you do next depends on the size of the study area.

In a very small community, the sampling rate must be relatively high (say, 1-in-5 households) to get an adequate sample size. Hence, travel time between households will be minimal. That means you may be able to walk through the entire community and mark every fifth household on a map. These are the households that interviewers should contact.

In a somewhat larger community, a lower sampling rate (say, 1 in 40) is more likely to produce an adequate sample size. To avoid high travel costs when the rate is low, sampling can be done in two stages. In the first stage, select a sample of blocks, giving every block an equal chance of selection. In a community with 400 blocks, you might select 40. In the second stage, within each block selected in stage 1, choose a sample of households large enough to produce the desired sample size. In this case, selecting every fourth household in each sampled block would produce the 1-in-40 sampling rate ($[40 \div 400] \times [1 \div 4] = 1 \div 40$). The result is that sample households are clustered in certain areas and interviewer travel costs are lower.

More efficient, but also more complicated, designs are described in Kish (1965).

So far, we have discussed sampling for face-to-face surveys in small geographic areas. As we mentioned, the sampling process becomes more complicated if the area that is being surveyed is not small. A large study area is only one of the conditions that warrant using more complex sample designs, to which we now turn our attention.

Why Use More Complicated Designs?

We expect that simple random or systematic sample designs will enable most of our readers to meet their survey objectives. By and large, these are manageable designs that can be implemented without substantial professional help. They are usually adequate for local surveys and even for some larger-area surveys conducted by mail or telephone.

Nonetheless, it is possible that some of our readers will need to use more complicated techniques. This is likely to be the case, regardless of survey method, if small subgroups within the target population are important to the research project. For example, consider a survey designed to study health status in a community of 5000 households of which 1000 are white and 4000 are black. Suppose the researchers are particularly interested in comparing health differences between blacks and whites. A 10 percent random or systematic sample would yield about 100 white and 400 black households. The subgroup of 100 white families may be too small for the intended analysis, either

What is a Weighted Sample?

Probability samples give every member of a population some known chance of being selected. In the simplest case, everyone has an *equal* chance of selection, for example, when every member of an organization is listed once and only once on a membership list. As long as everyone has an equal chance of being selected for a survey, generalizing from the sample to the population is straightforward. For example, if everyone has a 1-in-5 chance of selection, each element in the sample represents five elements in the population.

A problem arises when members of a population have *unequal* selection chances. This is usually the case, for example, with telephone directories. Most households are listed once but a few have more than one line and therefore have multiple listings. Those with more than one listing may be overrepresented in a sample because they have a greater chance of being selected than others.

Unequal chances of selection also occur when researchers use disproportionate sampling, a technique described in this chapter. Disproportionate sampling helps ensure that there are enough cases of various subgroups for the intended analysis.

A weighted sample is one that has been adjusted for unequal chances of selection. The purpose of weighting is to ensure that we can still generalize from the sample despite the fact that some subgroups may be over- or underrepresented.

Here is how weighting is done. In the example described in this section, white households had a 40 percent chance of selection and black households had a 10 percent chance. Technically, we say that the white households had a **sampling rate** of 0.4 and black households a rate of 0.1. The weight assigned to the different types of households is equal to:

$$= \frac{1}{\text{sampling rate}} \times \frac{\text{total sample size}}{\text{population size}}$$

$$= 2.5 \times \frac{800}{5000} = 0.4 \quad \text{for white households}$$

$$= 10 \times \frac{800}{5000} = 1.6 \quad \text{for black households}$$

We use the weights whenever we combine the two subsamples to create a composite picture of the whole community, in other words, whenever we make estimates about the population based on the sample.

because the sampling error on a sample of 100 is too large or because the subgroup will be broken down even further on the basis of income or another variable.

A practical solution to this problem is to sample at a higher rate in predominantly white neighborhoods than in black neighborhoods, say at 40 percent in one and 10 percent in the other. This would yield roughly 400 households of each race, enough to compare estimates of approximately equal precision.

It is likely that researchers working on such a project would want to put the black and white samples together and make estimates about the community as a whole. To do so, they would have to make an adjustment for having used two different sampling rates. The adjustment is called weighting. It can be done easily with most statistical software now on the market.

The technique we have just described is called **disproportionate** sampling. It can be used in conjunction with either random or systematic designs. However, it is important to note that calculating measures of sampling error in disproportionate samples can be very complicated. Readers interested in using the technique should definitely seek professional help before designing their sample.

Several times in this chapter we have mentioned another occasion when some of our readers might need professional help with sampling: when they are conducting a face-to-face survey in any area larger than a small community or neighborhood of several thousand households. In such a survey, both random and systematic sampling are inefficient and costly because interviewers must travel long distances between each interview.

A practical solution is to cluster interviews in small areas, such as census blocks. This is called **two-stage cluster sampling**. In the first stage, a sample of small areas is selected. In the second stage, a sample of respondents is chosen from the first-stage sample. The advantage of clustering like this is efficiency. Like disproportionate sampling, clustering can be used in conjunction with either random or systematic sampling.

We recommend getting professional help for surveys involving two-stage sampling. Determining the sampling rate at each stage and calculating sampling error within and among clusters is complicated and beyond the scope of our discussion.

Summary

Sampling is useful when the cost of getting completed questionnaires from everyone in the population is prohibitive. The number of people who need to be sampled depends on how much sampling error you can tolerate, the heterogeneousness (or variability) of the population in which you're interested, and how many subgroups within the sample you plan to analyze. Once you've made decisions on each of these issues, you can use a table to determine how large your sample should be.

An important part of sampling is compiling a list of the target population. Often, official sponsorship is critical to getting a good list. Unfortunately, however, few lists are perfect. Most have omissions, duplicates, and entries that are not pertinent to the study.

Successful mail surveys can't be done without good lists. For local telephone surveys, we recommend add-a-digit dialing, a relatively efficient sample design based on the telephone directory. For local face-to-face surveys, either a list-based or area-frame sample should be considered.

Simple random and systematic sampling are the simplest designs and are probably sufficient for most of our readers. More complicated techniques are necessary when small subgroups within the target population are important to the research project, regardless of the survey method. They are also essential in face-to-face surveys conducted in any area larger than a small neighborhood or community, and other surveys when efficiency is of great importance. For anything more complicated, we recommend that you seek professional help.

For more detail on simple random sampling, as well as on more complex methods, see:

Practical Sampling, by Gary T. Henry, Sage Publications, Newbury Park, CA, 1990.

Chapter 2, "Generating Sampling Pools," Chapter 3, "Processing Sampling Pools," and Chapter 4, "Selecting Respondents," in *Telephone Survey Methods,* by Paul J. Lavrakas. Sage Publications, Newbury Park, CA, 1987.

Chapter 3, "Sampling," in *Survey Research by Telephone,* 2nd Edition, by James H. Frey. Sage Publications, Newbury Park, CA, 1989.

Chapter 5, "The Logic of Survey Sampling" and Chapter 6, "Examples of Sample Designs," in *Survey Research Methods,* 2nd Edition, by Earl Babbie. Wadsworth, Belmont, CA, 1990.

Survey Sampling by Leslie Kish, John Wiley, New York, 1965.

Sampling Techniques, 3rd Edition, by William G. Cochrane, John Wiley, New York, 1977.

Appendix 5.A

How Sample Characteristics Influence the Size of Sampling Error

Sampling error occurs as a result of selecting a sample rather than surveying an entire population. When results from a sample survey are reported, they are often stated in the form "plus or minus" some number of dollars, percentage points, or whatever units are being used. For example, results of national presidential preference polls are often reported as being accurate plus or minus 3 percent.

The "plus or minus" reflects sampling error. To understand how sampling error is measured, let's put aside the other three kinds of error discussed in Chapter 2 and look at this one by itself.

In the eighteenth and nineteenth centuries, mathematicians and other scientists developed the theory that enables us to estimate sampling error. They explained that statistics based on samples drawn from the same population will vary from each other (and from the true population value) simply because of chance. The variation is **sampling error** and the measure used to estimate it is **standard error.**

The standard error of a sample proportion is calculated as follows:

$$se(p) = \sqrt{\frac{pq}{n}}$$

where $se(p)$ = standard error of a proportion

p and q = the proportions of our sample that do (p) and do not (q) have a particular characteristic

n = the number of elements in the sample.

Standard errors are used to estimate how close sample estimates are to a true population value, in other words, how precise the estimates are. (Remember, we have put aside the other three error sources for now.) Based on sample distribution theory, we are 95 percent confident that the true value is within two standard errors in either direction of the estimate. We are 68 percent confident that the true value is within one standard error in either direction. Another way of saying this is: Chances are 95 out of 100 that the estimate differs from the true value by less than two standard errors, and 68 out of 100 that it differs by less than one standard error.

For example, in a sample of 1259 registered voters, 30 percent favored a proposed bond to finance a new elementary school and 70 percent opposed it. The standard error of the proportion is

$$se(p) = \sqrt{\frac{(0.3)(0.7)}{1259}}$$

$$= .013, \quad \text{or } 1.3\%.$$

We are 95 percent confident that between 27.4 and 32.6 percent of *all* registered voters favor the bond (30 percent plus or minus 2 × 1.3 percent), and between 67.4 and 72.6 percent oppose it (70 percent plus or minus 2 × 1.3 percent). These ranges are called **confidence intervals.**

Three things about the sample affect the size of the sampling error and therefore how close sample estimates tend to be to true population values. First is the sampling procedure. Second is the variation within the sample with respect to the characteristic of interest. If our registered voter sample had been more uniform—say, 10 percent in favor and 90 percent opposed—the standard error would have been

$$se(p) = \sqrt{\frac{(0.1)(0.9)}{1259}}$$

$$= .008, \quad \text{or } 0.8\%.$$

Hence, the confidence interval would have been *smaller.* If the sample had been more varied—say, 50 percent in favor and 50 percent opposed—the standard error would have been 1.4 percent and the confidence interval would have been *larger.*

The third factor that affects the size of the sampling error is the size of the sample. A larger simple random sample results in a smaller sampling error. If we had a sample of 10,000 registered voters in which 30 percent favored the bond and 70 percent opposed it, our standard error would have been

$$se(p) = \sqrt{\frac{(0.3)(0.7)}{10{,}000}}$$

$$= .004, \quad \text{or } 0.4\%.$$

To sum up: Sampling design, variation, and sample size influence how much confidence we can have in our estimates. Not surprisingly, estimates based on more uniform and larger samples are likely to be more precise.

So far we have been looking at the standard errors of proportions. The same logic applies whenever we estimate a mean, which is one kind of average. The standard error of a mean is calculated as follows:

$$s_{\bar{x}} = \frac{s}{\sqrt{n}}$$

where $s_{\bar{x}}$ = standard error of the mean
s = standard deviation of the observations in the sample
n = number of elements in the sample.

The standard error of a sample mean is interpreted in the same way as the standard error of a sample proportion. With large enough samples, chances are 95 out of 100 that the sample mean is within two standard errors of the true population mean, in either direction. Chances are 68 out of 100 that the sample mean is within one standard error of the true population mean, in either direction.

For the sake of simplicity, we have been looking at standard error by itself. The other three error sources are equally important. However, since their magnitude often cannot be estimated, people tend to report sampling error as if it were the *only* kind that exists.

For more information, see Kachigan (1986) or Cochrane (1977).

Appendix 5.B

Regional Offices of the Census Bureau

Boston, MA
617/565-7200
Region: Maine, New Hampshire, Vermont, New York (excluding New York City, Nassau, Orange, Rockland, Suffolk, and Westchester counties), Massachusetts, Connecticut, Rhode Island

New York, NY
212/264-4730
Region: New York City, Nassau, Orange, Rockland, Suffolk, and Westchester counties, Puerto Rico

Philadelphia, PA
215/597-8313
Region: Pennsylvania, New Jersey, Delaware, Maryland

Charlotte, NC
704/371-6144
Region: District of Columbia, Virginia, North Carolina, South Carolina, Kentucky, Tennessee

Atlanta, GA
404/730-3833
Region: Georgia, Alabama, Florida

Kansas City, KS
913/236-3728
Region: Kansas, Oklahoma, Arkansas, Missouri, Iowa, Minnesota

Detroit, MI
313/226-7742
Region: Michigan, Ohio, West Virginia

Chicago, IL
312/353-0980
Region: Indiana, Illinois, Wisconsin

Denver, CO
303/969-7750
Region: North Dakota, South Dakota, Nebraska, Wyoming, Utah, Arizona, Colorado, New Mexico

Los Angeles, CA
818/904-6393
Region: California

Seattle, WA
206/728-5314
Region: Washington, Montana, Idaho, Oregon, Nevada, Hawaii, Alaska

6

Writing Good Questions

Do you agree with radical environmentalists who claim that 40,000 loggers should be put out of work to save 200 spotted owls?

Do you agree with timber company executives who argue that habitat for the few remaining spotted owls should be sacrificed so loggers can keep their jobs for the rest of their lives?

When it comes to collecting accurate and useful information about how people think and behave, survey questions like these are *not good*. In fact, they are downright useless. Depending on the context, "radical environmentalists" and "company executives" can be very emotional terms. Coupled with half-truths like jobs-for-owls and guaranteed lifetime employment, they can create problems with survey credibility, substantially biasing respondents. They yield ammunition for battle, not information that helps people make decisions.

Avoiding emotional and biased words is only part of writing good questions and, therefore, of minimizing measurement error. Other issues to consider are

- how specific the questions should be;
- whether the questions will produce credible information;
- whether respondents are able to answer the questions; and
- whether respondents will be willing to provide the information.

Turning Ideas into Useful Questions

In Chapter 3, we talked about building a foundation for survey research by identifying the problem that motivates a survey in the first place. This chapter is about translating the problem or idea into good questions—questions that respondents can understand and answer objectively. Social scientists have a technical word for this job; they call it **operationalizing**. It means setting up categories of events or phenomena that can be observed and measured.

Consider the example of a community action agency that did a survey to learn more about the causes of local poverty. They could not measure poverty without first defining it in terms of something that could be observed. The U.S. congressional Office of Management and Budget (OMB) has one way of defining poverty: Each year, it establishes a series of income thresholds for different family sizes and geographic locations. For example, in 1993 the poverty threshold for a four-person family living in one of the 48 contiguous states was $14,350. Families with income below the threshold are said to be poor. But the OMB definition is not sacred; people in other organizations and places may choose to operationalize poverty in different ways.

Consider another idea that could motivate doing a survey. Perhaps someone wants to know whether local residents are satisfied with their jobs. But "job satisfaction" has many dimensions, including adequate wages, a flexible schedule, and access to one's supervisor. Each dimension must be operationalized into things that can be measured, like wage rates, hours worked, workplace regulations, and scheduled meetings between employees and supervisors. The survey allows respondents to evaluate whether or not they are "satisfied" with each aspect of their job.

The key is this: To produce useful information, someone must take time to translate the ideas that motivate your survey into good questions. In the following discussion, we offer guidelines for achieving this objective, as well as many examples we hope are useful to you. Since we can't provide a foolproof recipe that would suit any situation, we encourage readers to make liberal use of the references listed at the end of this chapter and to get professional help when necessary.

What Kind of Information Are You Looking for?

The first step in writing a good question is to identify exactly what kind of information you want respondents to provide. Questions can usually be classified as asking for one of two types of information:

- What people do or what they are: their **behavior** or **attributes,** for example, whether they smoke cigarettes or were employed in the last 12 months.
- What people say they want or what they think is true: their **attitudes** or **beliefs,** for example, whether they favor a tax increase or believe that a 55-mile-per-hour speed limit saves lives.

Clarifying what kind of information you want is important because it is so easy to ask for one type of information when you really want another. For example, in a study designed to evaluate the local market for day care services, staff members at a social service agency must decide ahead of time whether they want to know about behavior or attitudes. If they want to know about behavior, such as how many families *now use* day care, then the following question is appropriate:

> While you are at work, is any child in your family regularly cared for, or not cared for, at a day care center or by another individual?

1 CARED FOR

2 NOT CARED FOR

Alternatively, if they want to know about attitudes, such as how residents feel about the adequacy of local day care services, a question like the following is appropriate:

In your opinion, are there enough, or not enough, slots for all the children in our community who need day care?

1 ENOUGH

2 NOT ENOUGH

Information from the first question can be used to describe current day care use, whereas information from the second can be used to describe how people view the adequacy of local services—two different pieces of information that cannot be substituted for each other.

A second reason for clarifying what kind of information you want is that questions of each type typically involve different degrees of measurement error. We explore why later in the chapter.

Which Kind of Question Structure Do You Want?

Virtually all questions that might be asked in a survey fit into one of four categories: open-ended, close-ended with ordered choices, close-ended with unordered choices, and partially close-ended. Each one tends to be better suited for obtaining certain types of information than others. We begin with the most difficult type of question to answer and analyze.

Open-ended

An example of an open-ended question is:

What should be done to improve this community?

Or alternatively:

In your opinion, why does our community have a poverty rate that is twice the national average?

Open-ended questions do not provide choices from which to select an answer. Instead, respondents must formulate an answer in their own words. This type of question takes the least amount of effort to write, but has several major drawbacks that should be taken seriously.

First, open-ended questions can be very demanding for respondents. People are asked to recall past experiences or discuss issues they may not have considered recently or at all. If making the interview quick and easy for respondents and those who analyze the data is important, you should probably avoid open-ended questions.

How Respondents Answered One Open-Ended Question

In a recent face-to-face survey of several hundred people, researchers asked the question, "What changes that have occurred in our community over the last 10 years do you like least?" Here are some of the answers they received:

More drugs in school
Unemployment for local residents
More robberies
High cost of living
Drugs among young people
More crime
High utilities
Outside firms taking profits
Businesses don't hire locals
Young people can't find jobs
Taxes are too high
Poverty has increased
Too many tourists
Racism has increased
Too many fast-food franchises
Need more stoplights
No affordable housing
Bypass is being planned
Low-wage jobs
Schools are overcrowded
Too many land speculators
Only jobs are seasonal
Bypass hasn't been built yet
No recreation for youth
No public transportation
Too much growth
Cultural values disappearing
More alcoholism
Outsiders buying land
More water pollution
Corrupt politics
Growth occurring too fast
Too much traffic
Poverty programs are being cut
Teachers don't get paid enough
City council is ineffective
Don't know my neighbors

Second, open-ended questions typically produce many different responses and only a few mentions of any one topic. The result is that a question like "What should be done to improve this community?" yields results such as 11 percent of local residents think schools should be improved, 9 percent think the roads need work, 8 percent think the mayor should be replaced, 7 percent think we need more parks, 5 percent think we have inadequate snow removal, 5 percent think the air is too dirty, and so on. By itself, information like this does little to guide public decision making.

Third, open-ended questions rarely provide accurate measurements or consistent, comparable information across the whole sample. If, with an open-ended format, 9 percent of the respondents say they think schools should be improved, it is likely that others would have agreed had they been provided with a stimulus to give the same answer. In addition, some respondents may give long, multipart answers to open-ended questions, while others comment only briefly. Comparing the two types of answers can be very difficult.

The fourth drawback is that open-ended questions require an enormous amount of time to prepare for later entry into a computer. Clerical staff must make at least two passes through some or all of the questionnaires, first to make a master list of responses, and second to **code** or assign numbers to each response so it can be entered into the computer. This procedure takes much more time than if choices are specified (and coded) ahead.

To summarize the disadvantages, open-ended questions are often difficult for respondents, may yield no more than a few mentions of any one topic, and are extremely difficult to code. However, they can be used successfully in certain circumstances. First, they can be used when researchers have little prior knowledge about a topic and therefore cannot specify response choices. In fact, open-ended questions are an excellent means of preparing to do a survey by exploring unknown subjects, such as someone might do in a focus group. Second, open-ended questions can be used when the main goal is to give survey respondents a chance to state strong opinions, vent frustrations, or let researchers know what has been overlooked. For example, at the end of a survey, respondents might be asked, "Is there anything else you would like to tell us about the subjects addressed in this questionnaire?"

Third, open-ended questions are sometimes helpful when they immediately follow a close-ended question (which we discuss below) and ask respondents to explain why they selected a particular answer. Their explanation may give researchers more insight regarding certain survey results.

Fourth, open-ended questions are desirable when people are being asked to estimate a routine behavior and are unlikely to know an exact number. For example:

How many hours per day do you watch TV?

Finally, open-ended questions are well-suited to situations in which a precise piece of information is needed and can be easily recalled without a list of answer choices, as in:

What state do you live in?

What make of automobile do you drive?

Very likely, respondents will not need a prepared list of choices to answer questions like these.

Plan Ahead!

Very few surveys are simple and small enough to tabulate questionnaires by hand. Instead, data are coded and entered into a computer so they can be analyzed later. Regardless of whether you do a mail, telephone, or face-to face survey, it is critical to PLAN AHEAD for data entry. That means designing your questionnaire so the people who code and enter the data can do their jobs efficiently. Before finalizing your questionnaire, read about data entry in Chapter 10. Avoid making costly mistakes that can't be remedied!

Close-ended with Ordered Choices

An example of a close-ended question with ordered choices is:

How do you feel about this statement? "This community needs more tennis courts." (Please circle the number of your response.)

1 STRONGLY DISAGREE
2 MILDLY DISAGREE
3 NEITHER AGREE NOR DISAGREE
4 MILDLY AGREE
5 STRONGLY AGREE

Or alternatively:

What is your present age? (Please circle the number of your response.)

1 UNDER 25 YEARS
2 26–35 YEARS
3 36–45 YEARS
4 46–55 YEARS
5 56–65 YEARS
6 OVER 65 YEARS

The distinguishing feature of this kind of question is that each choice represents a gradation of a single concept. In the examples above, the concepts are *how people feel about more tennis courts* and *current age*. For each question, the complete range of possible answers is provided. The respondent's job is to find the most appropriate place on the continuum for his or her answer.

Close-ended questions with ordered answer choices tend to be quite specific. Hence, they are less demanding for the respondent and much easier to code and analyze than open-ended questions.

Close-ended with Unordered Response Choices

An example of a close-ended question with unordered response choices is:

Which best describes the kind of building in which you live? (Please circle the number.)

1 A MOBILE HOME
2 A ONE-FAMILY HOUSE DETACHED FROM ANY OTHER

3 A ONE-FAMILY HOUSE ATTACHED TO AT LEAST ONE OTHER HOUSE
4 AN APARTMENT BUILDING

In close-ended questions such as these, answer choices are provided to respondents, but they don't fall on a continuum. People are asked to choose from among discrete, unordered categories. They must evaluate each choice and select the one that best reflects their situation.

Use the unordered format only when your knowledge of the subject allows you to list useful answer choices. For example, in a survey of college students (some of whom probably live in dormitories), you need a different list of housing options than those in the question above.

Close-ended questions with unordered response choices are usually more difficult than those with ordered answer choices. With ordered choices, people are asked to think only about where they fit on a continuum, for example, from strongly agree to strongly disagree. With unordered choices, they have to process more information. They must compare, for example, a "mobile home" with a "one-family house detached from any other." The job becomes progressively more difficult as more answer choices are added.

Close-ended questions with unordered response choices are often used to ask people to rank items, as in:

Who should be the most influential in deciding whether a downtown shopping mall is built in this community? Please rank the following individuals and organizations in order of their influence, starting with "1" for the person who should have the most influence.

	RANK (1 through 6)
The mayor	______
The city manager	______
The city council	______
The chamber of commerce	______
The downtown merchants	______
The general public	______

Or alternatively:

Who should be the most, second, and third most influential in deciding whether a downtown shopping mall is built in this community? (Put appropriate letter in each box.)

Should be:

☐ MOST	A	The mayor
	B	The city manager
☐ SECOND	C	The city council
	D	The chamber of commerce
☐ THIRD	E	The downtown merchants
	F	The general public

Questions in which respondents must rank unordered response choices are very difficult. Use them judiciously, especially in telephone surveys.

Partially Close-ended

An example of a partially close-ended question is:

> Which of the following areas of expenditure do you want to have the highest priority for improvement in this community?
>
> 1 STREETS AND ROADS
> 2 SEWAGE TREATMENT
> 3 PARKS
> 4 OTHER (PLEASE SPECIFY) __________

Such questions provide a compromise between the open- and the close-ended structures. Although answer choices are provided, respondents have the option of creating their own responses. The choices provided are almost always unordered.

Most respondents select one of the offered categories rather than developing their own, especially when the person conducting the survey has done a good job of identifying answer choices. Hence, this format rarely yields many additional responses. It does have the advantage of not forcing respondents into predefined boxes that don't fit the situation, and it occasionally generates new information.

Deciding Which Question Structure Is Most Useful

None of the four question structures outlined above is inherently best. Each has merits and is better suited to providing a particular kind of information. Knowledge of the four alternatives is useful in helping surveyors think through what they are really trying to find out.

For example, a state agency official wanted to know how farmers felt about problems facing U.S. agriculture in the 1990s. To help him, we used examples of the four different question structures.

We began by posing the open-ended question, "In your opinion, what problems face U.S. agriculture in the 1990s?" (See Question 1,

Figure 6.1 on next page.) Then we asked him to come up with the general answers he expected respondents to provide. He listed environmental, economic, and social problems. We then incorporated these three different kinds of problems into a close-ended question with ordered responses. Such a structure would allow him to ask how serious respondents think each individual problem is, as in Question 2, Figure 6.1. We then changed the question so that it was close-ended with unordered responses (as in Question 3, Figure 6.1) to see if he wanted respondents to compare environmental, economic, and social problems head-to-head.

Finally, to accommodate the concern that respondents might bring up an important problem that had been overlooked, we added an "other" answer choice. The resulting question was partially close-ended, as in Question 4, Figure 6.1.

Exercises like this help sort through what kind of question structure is most appropriate. We encourage you to try something similar in writing your own questions.

Does the Order of Response Choices Matter?

The order in which answer choices are listed can affect how people respond, but exactly how differs according to survey method. Evidence suggests that in mail surveys, people are somewhat more likely to choose from among the first categories listed. In telephone and face-to-face interviews, they are more likely to choose from among the last.

Category order effect, as it is called, may be most pronounced in questions with a long list of unordered categories, for example:

Which of the following reasons is *most likely* to make you look for a job with a new company?

1 LOW PAY
2 BAD RELATIONSHIP WITH CO-WORKERS
3 MISTRUST OF SUPERVISOR
4 INADEQUATE OFFICE SUPPORT
5 DISTANCE BETWEEN OFFICE AND HOME
6 LACK OF BENEFITS
7 LITTLE OR NO RECOGNITION FOR GOOD WORK
8 NO CHANCE FOR ADVANCEMENT

In a mail survey, respondents may read only the first few choices. However, in a telephone or face-to-face survey, interviewers are likely

Figure 6.1

A survey question structured four different ways

1. **Open-ended**

   In your opinion, what problems face U.S. agriculture in the 1990s?

   - Best for focus groups or other kinds of exploratory questioning.
   - Identifies range of answers that can be offered to respondents in more structured interviews later in the research process.
   - Rarely yields useful data for making reliable estimates about the percent of people with particular views or characteristics.

2. **Close-ended with ordered responses**

   Listed below are three problems some people believe exist for U.S. agriculture in the 1990s. In your opinion, how serious is each one? (Circle your answer.)

a.	ENVIRONMENTAL PROBLEMS	VERY	SOMEWHAT	NOT AT ALL
b.	ECONOMIC PROBLEMS	VERY	SOMEWHAT	NOT AT ALL
c.	SOCIAL PROBLEMS	VERY	SOMEWHAT	NOT AT ALL

   - Asks respondents to evaluate problems independent of each other.
   - Measures how serious respondents think each individual problem is.
   - Guides policy making by showing the extent to which one problem is viewed as more serious than others.

3. **Close-ended with unordered responses**

   In your opinion, which one of the following problems facing U.S. agriculture in the 1990s is the **MOST** serious? (Circle the number for your response.)

   1 ENVIRONMENTAL PROBLEMS
   2 ECONOMIC PROBLEMS
   3 SOCIAL PROBLEMS

   - Asks respondents to choose the single most important problem from a predefined list of alternatives (possibly those that policy makers have decided are legitimate or feasible targets for change).
   - Guides policy making toward action on the problem perceived as most important.

4. **Partially close-ended**

   In your opinion, which of the following problems facing U.S. agriculture in the 1990s is the **MOST** serious? (Circle the number for your response.)

   1 ENVIRONMENTAL PROBLEMS
   2 ECONOMIC PROBLEMS
   3 SOCIAL PROBLEMS
   4 OTHER (PLEASE SPECIFY)____________

   - Same as in type 3, but allows respondents freedom to identify important problems that researchers and policy makers have overlooked.

to read the last few choices before respondents have committed the first few to memory.

Category order effect can also be a problem in questions involving abstract ideas about which people haven't thought much, for example:

> Would you say that the lack of effective national leaders is more, about the same, or less of a problem than it was 10 years ago?
>
> 1 MORE
>
> 2 ABOUT THE SAME
>
> 3 LESS

On mail surveys, people tend to pick the first answer more often, while in a telephone or face-to-face interview, they tend to pick the last.

We cannot offer a foolproof solution to the problem of category order effect. Instead, we recommend that you be aware of how it might increase your measurement error. In telephone and face-to-face surveys, you may want to instruct interviewers to systematically vary the order in which they read answer choices. For example, for a question with eight choices, they would begin with the first choice in the first interview, the second in the second interview, and so on, through the eight choices.

In addition, try to keep your lists of answer choices from getting too long. To get information about abstract issues that are critical to the survey, use a series of questions rather than a single query.

Why Measuring Attitudes and Beliefs Requires Special Attention

Writing good questions about what people say they want or believe is true can be very difficult. To see why, we look first at questions about what people do or what they are. On most topics, asking questions about behavior and attributes is relatively simple and straightforward: People almost always give correct or true answers to questions about, for example, what time they set their alarm clock on weekday mornings, or about their age. If they make a mistake, it is usually unintentional and minor. If asked the same question again tomorrow, they will very likely give the same answer. In other words, asking about many personal attributes and behaviors produces very little measurement error.

The same cannot be said of questions on attitudes and beliefs. These we can measure much less precisely. Consider, for example, the spotted owl issue raised at the beginning of this chapter. Opinions about issues like this are very difficult to measure because they are often imprecise, change from day to day, and may not be well thought out in advance of the survey.

Consider another example:

To what extent do you agree or disagree with the statement, "Parents should punish children who disobey?"

1 STRONGLY DISAGREE
2 SOMEWHAT DISAGREE
3 NEITHER DISAGREE OR AGREE
4 SOMEWHAT AGREE
5 STRONGLY AGREE

When asked a question like this, people typically think awhile and then answer a little uncertainly. If asked the same question tomorrow, they may give a different response.

People do not "possess" attitudes and beliefs in the same way they "possess" attributes like age and gender. Hence, answers to questions like the ones on spotted owls and whether children should be punished involve a high potential for measurement error. That doesn't mean questions on attitudes and beliefs should never be asked, but it does mean they require special attention.

Psychologists and sociologists often use a **scaling** technique to improve their ability to measure attitudes and beliefs. This means that instead of writing just one question about spotted owls, for example, they write a series of questions like those shown in Figure 6.2. Then they look for patterns in how people answer questions in the series and use statistical techniques to combine answers into a single numerical estimate. The advantage of using such a series of questions is that the complex issues of environmental protection and economic security are not oversimplified. The scaling technique thus gives researchers a way to collapse answers to the whole series into one indicator on how people really think about an issue.

Scaling is a well-established and useful technique that is explained in textbooks like those listed at the end of this chapter. However, it is probably impractical for people who do the kind of surveys we discuss here, not least because it requires sophisticated skills. Furthermore, the results are hard and sometimes almost embarrassing to explain to citizens and public officials. A statement like "54 percent of local residents are satisfied with community services" has an obvious meaning. In contrast, saying that "60 percent of local residents scored 3.6 on a satisfaction scale of 1 to 5, based on combined responses from eight questions" has a less-than-obvious meaning for most people.

To those who are not inclined to use scaling, for whatever reason, we offer this advice. First, with a passion, avoid using extremely abstract questions that will inevitably lead to measurement error.

Figure 6.2
Using a series of questions to get opinions on a single topic

A range of opinions exists about plans to protect spotted owls and about how these plans might affect the timber industry in the Pacific Northwest. Please indicate the extent to which you agree or disagree with each of the following statements by using the scale below:

SA — STRONGLY AGREE
MA — MILDLY AGREE
U — UNDECIDED OR UNSURE
MD — MILDLY DISAGREE
SD — STRONGLY DISAGREE

Statements	Indicate the extent to which you agree or disagree with each statement. (Please circle your answer.)				
1 It is very important that the spotted owl be protected from extinction, regardless of the impact on employment in the timber industry	SA	MA	U	MD	SD
2 Saving loggers' jobs is as important as preserving the spotted owl from extinction	SA	MA	U	MD	SD
3 It is more important to save loggers' jobs than to preserve the spotted owl from extinction	SA	MA	U	MD	SD
4 Loggers who lose their jobs because of spotted owl protection plans should be retrained with publicly funded programs	SA	MA	U	MD	SD
5 The federal government should reach a compromise with timber companies so that loggers can keep their jobs and the spotted owl population will survive	SA	MA	U	MD	SD

Second, on abstract issues that your survey simply must address, use different kinds of question structures so that you rely less on any single question.

Consider a survey recently conducted for a group of county commissioners. They wanted to know how local residents felt about the way housing developments had been spreading into prime agricultural land. For this survey, we did *not* ask respondents to tell us their opinions on the following statements:

> Saving agricultural land for future generations is virtuous.
>
> It is a good idea to protect green space around cities by protecting it from encroaching housing developments.

Instead, we *did* ask for opinions on questions like the following:

> Should the commissioners in Grant County take action or not take action that would discourage housing development on agricultural land surrounding Centerville?
>
> Which is more important—protecting people's freedom to build homes in rural areas of Grant County or preserving land that is now being farmed?
>
> Should county commissioners discourage or not discourage development of agricultural land surrounding Centerville, even if it increases land prices in other parts of the county?

After completing the survey, we showed the results to the county commissioners. They were able to compare and think about the consistency of people's answers and thus make their own judgments about public opinion on the issue.

Encouraging Thoughtful Answers

Sometimes measurement error occurs because people answer survey questions quickly and without thinking. This is especially true when they are asked to recall specific events over a long time period, as in "How many times have you gone to a doctor in some other community during the last nine months?" Mail surveys seem to present the most serious problem in this regard, because some respondents go through their questionnaires so quickly.

To avoid measurement error like this, we suggest asking a series of questions that encourages respondents to recall the particular event or time period before they answer the really important part of the question. For example, instead of asking point-blank, "Did you wear or not wear your seatbelt the last time you were an automobile passenger?," consider a series like this:

> When was the last time you rode in a car as a passenger rather than as a driver? Was it:
>
> 1 TODAY
> 2 YESTERDAY
> 3 BEFORE YESTERDAY
>
> How long was the trip? Was it:
>
> 1 LESS THAN A MILE
> 2 1–2 MILES
> 3 MORE THAN 2 MILES

Did you wear your seatbelt:

1 THE ENTIRE TIME THE CAR WAS MOVING
2 PART OF THE TIME THE CAR WAS MOVING
3 NOT AT ALL

This is called a **cognitive** design. When used throughout a survey, it makes for a very long questionnaire. However, it is effective for some particularly important questions that require accurate answers.

How to Overcome Common Wording Problems

The final step in writing good questions is to consider the exact wording. No one has ever succeeded in compiling a perfect list of rules for writing good questions, although many have tried. The reason such a list is hard to make is that principles that seem sensible often get in each other's way and only confuse us. For example,

- Be specific.
- Use simple words.
- Don't be vague.
- Keep it short.
- Don't talk down to respondents.
- Don't be too specific.

"Use simple words" and "Don't talk down to respondents" are both reasonable suggestions, but they get in the way of "Keep it short." Likewise, the combination of "Don't be vague" and "Don't be too specific" puts even the most experienced surveyor in a quandary.

The real impediment to making a foolproof list is that survey researchers do not write questions in the abstract. Instead, they write for a particular population and purpose, and in the context of other questions in the survey.

Since no one can provide a set of absolute rules, we suggest using the references listed at the end of this chapter; *consider* using questionnaires that have worked in other surveys on a similar topic. We also suggest studying the wording problems illustrated in Appendix 6.A. To illustrate each problem, we provide a question that is likely to produce inaccurate or unusable information. We also suggest a way to revise the question and then briefly discuss the wording problem that is illustrated.

Summary

Writing good questions usually takes more than one, two, or even three sittings. A friend once pointed to a full wastebasket when we asked how her questionnaire was coming along. Only many days later, when we inquired again, did she hand us several pages of neatly typed questions.

The reason that writing questions usually takes time and many attempts is that so many things have to be taken into consideration all at once—deciding what new information is needed, how to structure questions, and whether people can accurately answer what is asked—all while avoiding a variety of wording problems.

Fundamentally, writing good questions means minimizing measurement error. However, once questions are written, this objective is only partially achieved. Still, the questions must be ordered and arranged on the pages of a printed questionnaire, the task to which we turn our attention next.

For more detail on writing good questions, see:

Asking Questions: A Practical Guide to Questionnaire Design, by Seymour Sudman and Norman M. Bradburn, Jossey Bass, San Francisco, CA, 1982.

Chapter 3, "Writing Questions: Some General Principles," in *Mail and Telephone Surveys: The Total Design Method,* by Don A. Dillman. Wiley-Interscience, New York, 1978.

Appendix 6.A

Common Wording Problems and Possible Solutions

Wording Problem	Possible Revision
Vaguely worded questions and responses	
Do you actively support, support, or not support expanded public transportation services in your community?	A proposal has been made to increase the number of public bus routes so that residents will not have to travel as far to catch a bus. Would you be willing or not willing to pay an increase of 1 percent in the city gas tax to finance more bus routes?
1 ACTIVELY SUPPORT	1 WILLING
2 SUPPORT	2 NOT WILLING
3 NOT SUPPORT	3 UNDECIDED

KEY POINT

Vaguely worded questions and responses produce useless information. The question in the left-hand column gives decision makers no information they can act on because they can't tell what "support" means.

Wording Problem

Abbreviations or jargon that respondents may not understand

Do you believe LISA research should or should not be funded by the USDA?

1 SHOULD
2 SHOULD NOT
3 UNDECIDED OR UNSURE

Possible Revision

The U.S. Department of Agriculture (USDA) recently began a research program on "low-input sustainable agriculture" (LISA). The program is primarily designed to investigate farming techniques that use reduced levels of agri-chemicals. Do you favor or not favor having a USDA research program on LISA?

1 FAVOR
2 DO NOT FAVOR
3 UNDECIDED OR UNSURE

KEY POINT

Undefined abbreviations and jargon should not be used. The exception is when researchers are surveying a particular group of people who use the terms regularly and will definitely not be confused.

Wording Problem

Too much precision

How many times, if any, did you, yourself, buy gasoline last year? ________

Possible Revision

How many times, if any, did you yourself buy gasoline last year?

1 NEVER
2 1–12
3 13–24
4 24–52
5 MORE THAN 52

KEY POINT

Sometimes too much precision makes questions nearly impossible to answer and may provide more detailed data than the researcher really needs. Broad categories make the respondent's job easier.

Wording Problem

Bias from slanted introduction

More Americans exercise regularly now than they did 10 years ago. Do you exercise—such as bike, walk, or swim—regularly, or do you not exercise regularly?

1 DO EXERCISE REGULARLY
2 DO NOT EXERCISE REGULARLY

Possible Revision

Do you exercise—such as bike, walk, or swim—regularly, or do you not exercise regularly?

1 DO EXERCISE REGULARLY
2 DO NOT EXERCISE REGULARLY

KEY POINT

Bias is created when questions make it seem as though *everyone* engages in a particular activity and therefore the respondent should too.

Wording Problem

Bias from unequal comparison

Who do you feel is most responsible for the high cost of U.S. automobiles?

1 AUTOWORKERS
2 AUTO COMPANY EXECUTIVES
3 CONSUMERS

Possible Revision

Who do you feel is most responsible for the high cost of U.S. automobiles?

1 WORKERS WHO PRODUCE AUTOS
2 AUTO COMPANY EXECUTIVES WHO MANAGE MANUFACTURING PLANTS
3 CONSUMERS WHO BUY AUTOS

KEY POINT

It's easy to blame a small group of privileged people. Responses in the right column are intended to make the comparison more equal and eliminate bias.

Wording Problem

Bias from unbalanced response choices

The projected annual cost of agricultural legislation passed by the U.S. Congress in 1990 was roughly $10 billion. Do you believe this amount is:

1 TOO LITTLE
2 ABOUT RIGHT
3 SLIGHTLY TOO HIGH
4 MODERATELY TOO HIGH
5 FAR TOO HIGH

Possible Revision

The projected annual cost of agricultural legislation passed by the U.S. Congress in 1990 was roughly $10 billion. Do you believe this amount is:

1 FAR TOO LOW
2 SLIGHTLY TOO LOW
3 ABOUT RIGHT
4 SLIGHTLY TOO HIGH
5 FAR TOO HIGH

KEY POINT

Bias can be created when responses are weighted in one direction. The list in the left column has an unequal number of positive and negative choices.

Wording Problem

Bias from the tone of the question

Do you agree that garbage from big cities should be dumped in rural landfills to provide small communities with what will likely be modestly higher tax revenues?

1 NO
2 YES
3 UNSURE OR UNDECIDED

Possible Revision

Some people suggest that one way of increasing revenues in small communities is to allow waste from urban areas to be transported to rural landfills. Under this plan, landfill owners would pay property taxes to local governments in the communities with the landfills. In your opinion, should rural communities consider or not consider this proposal?

1 CONSIDER
2 NOT CONSIDER
3 UNSURE OR UNDECIDED

KEY POINT

A subjective tone introduces bias. The question in the left column uses words that might slant the respondent's answers or discredit the survey's objectivity, while the one on the right does not.

Wording Problem

Objections to providing income information

How much profit did your business earn in 1990?

_____________ DOLLARS

Possible Revision

How much profit did your business earn in 1990?

1 COMPANY LOST MONEY
2 ZERO TO $9,999
3 $10,000 TO $19,999
4 $20,000 TO $39,000
5 MORE THAN $40,000

KEY POINT

Many people will not answer personal questions, such as those about income. One method that *sometimes* works to overcome this problem is using broad response categories.

Wording Problem

Objectionable statements

"Single-parent households are a direct result of the way the U.S. welfare system is set up." Do you agree or disagree with this statement?

1 AGREE
2 DISAGREE
3 UNSURE OR UNDECIDED

Possible Revision

I would like to ask you about the relationship between public assistance programs (sometimes called *welfare*) and family stability. Here are various opinions we have heard people give on the topic. We would like to know whether you agree or disagree with each one.

"Welfare enables low-income, single parents to stay home and be available to meet their children's needs."

1 AGREE
2 DISAGREE
3 UNSURE OR UNDECIDED

"Single-parent households are a direct result of the way the U.S. welfare system is set up."

1 AGREE
2 DISAGREE
3 UNSURE OR UNDECIDED

Comment _____________________

KEY POINT

Establishing a context to soften the impact is another way of overcoming reluctance to answering objectionable questions. The question in the left column might be offensive to some people. Placing it within a series of statements removes some of the stigma.

Wording Problem

Questions that are too difficult for respondents

What percentage of your weekly grocery bill is spent on dairy products?

_______ PERCENT

Possible Revision

About how much money do you spend each week on the following items?

MILK	$_____.00
CHEESE	$_____.00
COTTAGE CHEESE	$_____.00
SOUR CREAM	$_____.00
OTHER DAIRY PRODUCTS	$_____.00

About how much do you spend on all groceries each week?

_______ DOLLARS

KEY POINT

Figuring percentages without a calculator is difficult and, even with calculators, errors are common. If possible, request information that will allow you to perform the calculation after the interview is over.

Wording Problem

Double-barreled questions

Do you favor protecting U.S. textile manufacturers from foreign competition but not U.S. farmers?

1 YES
2 NO
3 UNSURE OR UNDECIDED

Possible Revision

Do you favor or not favor protecting U.S. textile manufacturers from foreign competition?

1 FAVOR
2 NOT FAVOR
3 UNSURE OR UNDECIDED

Do you favor or not favor protecting U.S. farmers from foreign competition?

1 FAVOR
2 NOT FAVOR
3 UNSURE OR UNDECIDED

KEY POINT

Double-barreled questions produce ambiguous answers. Split the question into two parts so respondents can answer one part at a time.

Wording Problem

Answers are not mutually exclusive

How did you first hear about the proposed sales tax change?

1 FROM A FRIEND OR RELATIVE
2 AT A MEETING OF AN ORGANIZATION TO WHICH I BELONG
3 AT WORK
4 FROM MY SPOUSE
5 FROM THE TELEVISION, RADIO, OR NEWSPAPER

Possible Revision

From whom or what did you first hear about the proposed sales tax change?

1 FROM A FRIEND OR RELATIVE
2 FROM MY SPOUSE
3 OVER THE TELEVISION OR RADIO
4 FROM THE NEWSPAPER

Where were you when you first heard about the proposed sales tax change?

1 AT A MEETING OF AN ORGANIZATION TO WHICH I BELONG
2 AT HOME
3 AT WORK

KEY POINT

If you want only one response to a question, make sure the choices are mutually exclusive. Otherwise, you won't get complete and accurate information.

Wording Problem

Too much knowledge is assumed on the part of the respondent

Do you tend to agree or disagree with the U.S. Department of Agriculture's new regulations on labeling criteria for organic produce?

1 AGREE
2 DISAGREE
3 UNSURE OR UNDECIDED

Possible Revision

The Department of Agriculture has recently issued new regulations regarding criteria for using the "Organic" label on fresh produce. Were you aware or not aware that the agency had issued new regulations?

1 AWARE
2 NOT AWARE

If aware, please describe these regulations in your own words.

____

____

____

Are these regulations acceptable or not acceptable to you, personally?

1 ACCEPTABLE
2 NOT ACCEPTABLE
3 UNSURE OR UNDECIDED

KEY POINT

Do not assume the respondents know enough to answer your questions. Verify what they know first, and then proceed. The same admonition applies to behavior. If you ask about what kind of restaurants they eat at, first make sure that going to restaurants is something they do.

Wording Problem

Inaccurate statements

One of the most important revenue sources for school districts is the local sales tax. In the future, should our community's schools rely more heavily, about the same, or less heavily on sales taxes?

1 MORE HEAVILY
2 ABOUT THE SAME
3 LESS HEAVILY

Possible Revision

One of the most important revenue sources for school districts is the *property* tax. Some people suggest communities should decrease their dependence on property taxes. In the future, should our community's schools rely more heavily, about the same, or less heavily on *sales* taxes?

1 MORE HEAVILY
2 ABOUT THE SAME
3 LESS HEAVILY

KEY POINT

Inaccurate statements quickly destroy a survey's credibility by discrediting the researchers' intelligence. Make sure you know what you're talking about.

Wording Problem

Inappropriate time references

How many hours a day did you work last week?

__________ HOURS

Possible Revision

How many hours a day did you work during the first week of June 1991?

__________ HOURS

OR

On average, how many hours a day do you usually work?

__________ HOURS

KEY POINT

Surveys often take weeks or months to complete. "Last week" refers to a different time period when interviews are conducted weeks apart. In a survey of seasonal workers, for example, the question in the left column will not produce the same information for all respondents.

Wording Problem

Responses that can't be compared with existing information

How many years of school have you completed?

1 0 TO 8
2 9–10
3 11–12
4 MORE THAN 12

Possible Revision

How many years of school have you completed?

1 0 TO 8
2 9–11
3 12
4 MORE THAN 12

KEY POINT

An excellent way to test how well your sample represents the population in which you are interested is to compare survey data with published statistics. But that can't be done unless you ask for the information in exactly the same way as the published data source. Answer choices in the right column are the same ones published by the U.S. Census, while those on the left are not.

Wording Problem

Cryptic questions that may be misunderstood by respondents

Q-1 Number of years employed in present job

_____________ YEARS

Q-2 Your occupation

____

Possible Revision

Q-1 How many years have you been employed in your present job?

_____________ YEARS

Q-2 What is your present occupation?

____

KEY POINT

Incomplete sentences don't substitute for good questions. In the example on the left, respondents are likely to confuse the information requested in Question 2 with that in Question 1. Hence, they may answer Question 2 as if it read "How many years have you been in your present occupation?"

7

Questionnaire Design

Sometimes people make a list of what they want to find out and then assume that producing a questionnaire is simply a matter of writing questions to fill the page. Ordinarily, that's not the case.

> A city council worked hard to identify a list of possible goals for developing the local economy. They showed their list to others in the office, and because the goal choices were familiar, those people found it easy to rank which goal was most important. However, when the council read the list over the telephone in a survey of local residents, virtually no one could remember the choices long enough to provide reasonable answers.

> In another community, a housing task force drew up a list of questions to be asked of local residents. First on their list was the question most critical to their task. It was about income. They hired someone to do a mail survey for them, and, much to their surprise, the first item on the printed questionnaire turned out to be, "How much was your total family income in 1987?" Many respondents were unwilling to answer such a personal question when it was the first thing they were asked; thus the overall response rate suffered.

This chapter is about turning a laundry list of questions into a finished questionnaire to which people can respond easily and accurately. Specifically, we are concerned with reducing both nonresponse *and* measurement errors that come from hard-to- complete questionnaires.

Well-designed questionnaires take time to put together, but their payoff is enormous. People are *willing* to respond to attractive questionnaires, so nonresponse error is minimized; and they are able to do so *accurately*, so measurement error is less of a problem as well.

Good questionnaires make the task of responding as easy as possible, or as the professionals say, they minimize **respondent burden**. This means decreasing the time actually required to complete the questionnaire as well as the time respondents *think* is required. It also means making questions easy to answer rather than confusing and difficult. And last, it means showing respect for respondents by making sure they will not be embarrassed by not understanding what is expected of them.

Perhaps the biggest barrier to writing a good questionnaire is the need to realize that mail and interview questionnaires look, sound,

Plan Ahead

Very few surveys are simple and small enough to tabulate questionnaires by hand. Instead, data are coded and entered into a computer so they can be analyzed later. Regardless of whether you do a mail, telephone, or face-to-face survey, it is critical to *plan ahead* for data entry. That means designing your questionnaire so the people who code and enter the data can do their jobs efficiently. Before finalizing your questionnaire, read about data entry in Chapter 10. Avoid making costly mistakes that cannot be remedied!

and—when held in one's hands—may even feel different. Respondents only *hear* the words in a telephone questionnaire and obviously are not at all influenced by what the questionnaire looks like. In contrast, respondents *see* a mail questionnaire, usually before they have any idea of what it is about. The initial impression—words or pictures on the cover, size, neatness, and other physical characteristics—may either motivate or discourage people from starting to fill it out.

Additionally, when questions are written for mail surveys, the answers are often displayed below the question and are not incorporated into the set of words read to respondents, that is, the **main stem** of the question. In contrast, imagine what would happen if a telephone interviewer read the question, "Which of the following categories best describes why you chose to drop out of school?" and then fell silent, waiting for a response.

Mail, telephone, and face-to-face questionnaires each rely on a different way of communicating. Mail questionnaires are visual, telephone interviews depend on what one hears, and face-to-face interviews can utilize all forms of communication. What makes a good survey form for one does not work for another. It is to these issues that we turn next.

Mail Questionnaires

Mail questionnaires are unique because they stand on their own. No interviewer is present to convince respondents that they should fill out the questionnaire. Therefore, motivating people to respond is one of the important goals in designing mail questionnaires.

What makes people respond to mail surveys? The answer is that they respond when they think the survey is worth their time—that it will not be too difficult or take too long. This is the challenge of designing good mail questionnaires, and meeting it depends on knowing how people respond to visual signals and how they use written instructions.

Imagine that you have just bought a new VCR and are trying to make it work. Do you study the entire owner's manual, reading it word-by-word and examining all the diagrams? Or do you simply forge ahead, hooking it up in a way that seems logical and then try to tape a program, referring to the manual only when you run into problems?

Good mail questionnaires are designed for those who forge ahead! Most people don't read cluttered pages, full of fine print. They want consistent instructions, only as many as necessary, laid out clearly and exactly where they apply. The visual impact is critical.

At its best, the U.S. interstate highway system is a good example of how instructions can be given with visual clarity. When highway engineers do their job right, drivers don't have to stop to figure out where they are supposed to go. Entrance ramps are identified by green signs with arrows. Overhead signs give information about the distance to upcoming exits and are placed above lanes from which the exits can be accessed. Signs like **DES MOINES: NEXT 5 EXITS** notify drivers about what is ahead but aren't loaded down with details about which exit goes where. Blue and white signs along the side of the highway use symbols for gas stations and restaurants to indicate services available at upcoming exits. In other words, information is provided in a graphically consistent way and exactly where it is needed.

We strive to accomplish the same with mail questionnaires. We start with only essential instructions instead of "blocking the entrance" with extraneous wording right up front. We don't begin with "Please answer all the questions!" or with directions that will not be needed until later: for example, "If you have a college degree, you will be asked to skip certain items in this questionnaire." Instead, we use the front of the questionnaire to attract people's interest and to convince them that the survey is worth their time. That's why format and printing are the first specific design features to which we turn our attention.

Format and Printing

Several years ago, one of us developed a set of exact instructions on how to print questionnaires that get people's attention and obtain high response rates. Among the recommendations were the following:

- Print the questionnaire as a booklet, with dimensions of precisely $6\frac{1}{8}$ by $8\frac{1}{4}$ inches.
- Make the booklet from legal-size paper that you have trimmed to $8\frac{1}{4}$ by $12\frac{1}{2}$ inches and then folded in the middle.
- Type each page with 12-point elite type on regular-size paper ($8\frac{1}{2}$ by 11 inches) and then photographically reduce it to 79 percent of the original.
- Print the booklet on white or off-white 16-pound bond paper.

Following these recommendations resulted in a professional-looking booklet that could be folded into a regular or slightly smaller envelope. If the questionnaire contained 12 pages (including a front and back cover), it could be mailed for one first-class stamp.

Advances in desktop publishing have made these instructions outmoded, and in retrospect they were more rigid than they needed to be. Still, the general principle makes sense for today's mail questionnaires. A booklet format with questions printed on both sides of each inside page is an efficient and professional-looking printing method. However, we have found that readily available—and untrimmed—legal-size paper works very well in our questionnaires, and the type size can be varied according to respondents' needs. We might use type larger than 12 point, for example, in a survey of the elderly.

The Front Cover. Simply put, use the front cover to make the questionnaire look interesting. It offers a perfect opportunity to spark people's curiosity. This is *not* the place to put a cover letter, detailed instructions, mailing labels, promotional information about the sponsoring agency, or the first question.

Besides stimulating interest, an attractive cover helps set a particular survey apart from others and also conveys the idea that someone has worked hard to develop the questionnaire. Past research has found

Questionnaire Titles, Some Good and Some Bad

Good Titles	Bad Titles
Honest: The Washington Seat Belt Law: An Assessment of Citizen Experiences and Opinions	*Deceptive:* Your Last Chance to Influence Washington State Seat Belt Laws!
Interesting: Employee Training and Retraining in the 1990's: A Survey of Arizona Employers	*Boring:* Training Issues Survey
Simple: Business Concerns in Milwaukee	*Academic:* A Multivariate Analysis of Factors Affecting Business Decisions in Milwaukee
Neutral: Meeting Future Electricity Needs in the Pacific Northwest	*Biased:* Must We Build Yet Another Dam?

that when people are sent a reminder postcard, they remember unusual covers and can retrieve the questionnaire from wherever they have left it.

A well-designed cover includes only four items: an informative title that motivates respondents to open the questionnaire; a graphic design or illustration that helps identify the survey; the name of the study's sponsor; and the address where the questionnaire is to be returned. The front cover may also be a place to list a toll-free telephone number for anyone who has questions or needs help.

The **study title** should be clear and to the point, conveying as simply as possible what the survey is about. Subtitles are often useful, especially when they provide more detail or tell who is being surveyed. Good titles are memorable, but they should not threaten, mislead, suggest bias, or sound academic.

A **graphic design** or **illustration** adds interest to the front cover. It can be simple or complex and need not symbolize the study topic. For example, it may be abstract or be an outline map of the state in which the study is being done.

We recommend that cover graphics be neutral and not suggest bias or controversy. For example, on the cover of a mail questionnaire about gun control in Wyoming, an outline map of the state would be better than a picture of a hunter. For a study of driving habits, the picture on the cover should not be of passengers without seat belts.

Some people put several different graphics on a questionnaire cover to show a balance between opposing issues. However, because many respondents focus only on what they *want* to see, we prefer a neutral design. For example, for a survey on the relative importance of economic development and environmental preservation, an abstract symbol would be better than a drawing of a pristine wilderness area and another of a new shopping mall.

Finally, the last two items on the front cover are the **name of the study's sponsor** and the **return address** to which the completed questionnaire should be mailed. This information is important—if a respondent loses the reply envelope and cover letter that come with the questionnaire, he or she needs to know where to return the completed form. To maintain an objective and businesslike approach, we do not add the researcher's name to the return address.

Figure 7.1 shows a front cover design that makes a good first impression. It is abstract, neutral, uncluttered, and makes no personal connection with the researchers. Figure 7.2 shows a design that makes a bad first impression. It is cluttered, hard to read, may suggest bias, and includes the researchers' signatures.

Figure 7.1
A good design for the front cover of a mail questionnaire

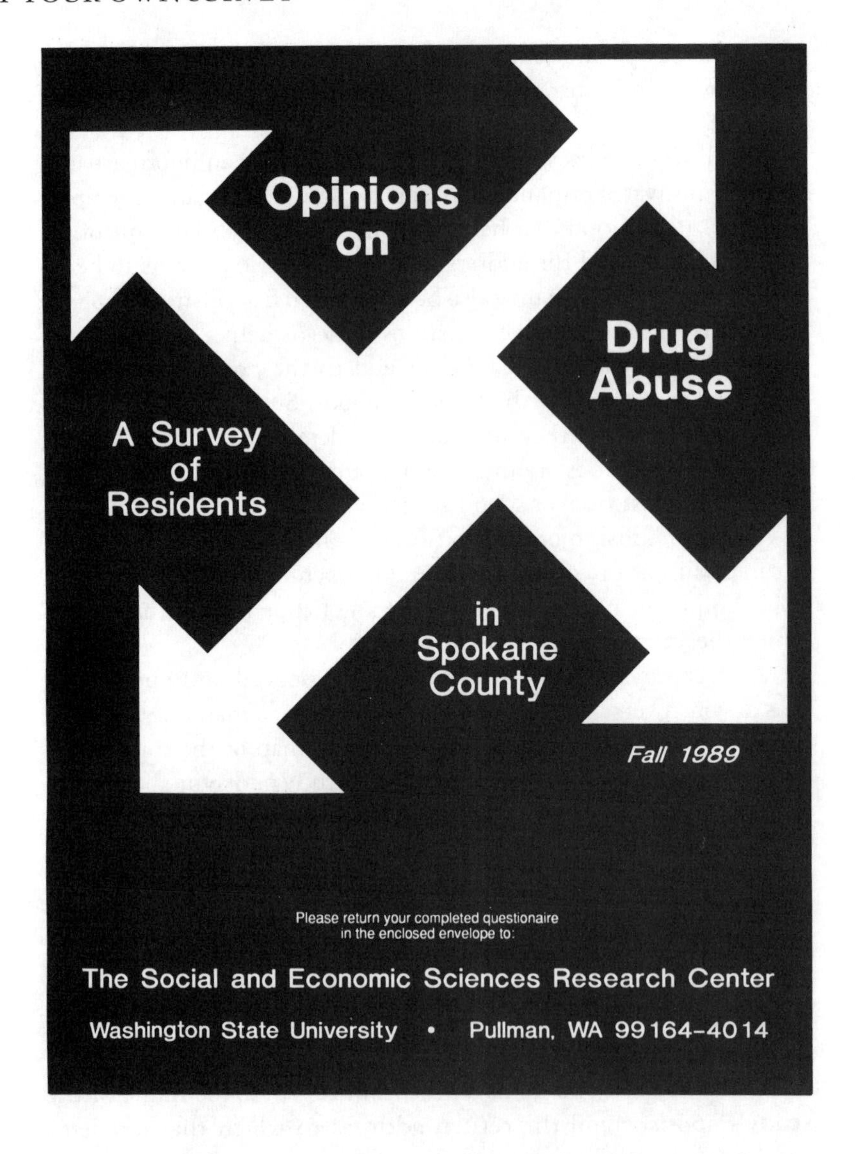

The Back Cover. On the back cover should appear an invitation to make comments, plenty of blank space where the respondent can write, and at the bottom of the page, a thank-you for completing and returning the questionnaire (Figure 7.3). We always ask for comments because they invariably teach us something useful. Sometimes people point out questions they couldn't understand or that didn't apply to them. Other times, they offer new insights into the subject of the survey. And always, giving people a chance to make voluntary, impromptu remarks promotes goodwill. If we don't specifically

SURVEY OF NATURAL RESOURCE AND FORESTRY ISSUES

In recent years there has been a lot of discussion and controversy concerning natural resource and forestry issues in our state. In this survey we want to find out what you think about these and other related matters. The study is being conducted by researchers at Central State University. Your name has been drawn from a telephone directory as part of a random sample of residences. Your participation in this survey is completely VOLUNTARY; however, in order to gather a fair impression of how citizens feel about these issues, it is important that as many people as possible respond to the survey. Your answers will be kept COMPLETELY CONFIDENTIAL. The identification number at the bottom of the page is only for mailing purposes; no record of these numbers will be retained once the survey is completed. If you would like a copy of the results, please include a note with your addess and "COPY OF THE RESULTS REQUESTED" written on it. Please do not record this information on the questionnaire itself. Thank you for your time and effort.

Respectively,

Christine Phelps	Steven Hill	Paul Gibson
Political Science	Philosophy	Forest Resources
Central State Univ.	Central State Univ.	Central State Univ.
(599) 299-6703	(599) 299-6143	(599) 299-6167

ID#____________________
(For mailing purposes only.)

Figure 7.2
A bad design for the front cover of a mail questionnaire

encourage comments on a mail questionnaire, we miss what would likely come as a matter of course during a telephone or face-to-face interview.

As we discuss later, questions to which people are more likely to object are best placed near the end of a questionnaire. However, we recommend against letting these questions spill over onto the back cover where they may be seen before anything else. Sensitive questions about income, political party, or religion, for example, are normally placed near the end of the questionnaire but shouldn't be allowed to dictate the respondents' first impression of the study.

In some instances, as when mail questionnaires are personally handed to people who are visiting a park or museum, we reserve the

Figure 7.3
A good design for the back cover of a mail questionnaire

Your comments will be appreciated, either here or in a separate envelope.

Thank you for your help.

Please return your completed questionnaire in the enclosed envelope to:
SOCIAL & ECONOMIC SCIENCES RESEARCH CENTER
Washington State University
Pullman, WA 99164-4014

back page for a postage stamp and return address; any voluntary comments are asked for on the last inside page of the questionnaire. The reason for making the questionnaire self-contained is that return envelopes are too easily lost. A self-sticking tab to seal the questionnaire for mailing is also used.

Figure 7.3 shows a back cover design that includes ample space for comments and doesn't distract from the front cover. Figure 7.4 shows a design that has very limited space for comments and includes sensitive questions that may give an unfavorable first impression.

36. Which of the following categories best describes your yearly total household income?
Circle one answer.

1) less than $ 5,000
2) $ 5,000 - 9,999
3) $10,000 - 14,999
4) $15,000 - 24,999
5) $25,000 - 49,999
6) $50,000 or more

37. Which of the following categories best describes your racial or ethnic identification?
Circle one answer.

1) BLACK
2) CHICANO OR HISPANIC
3) NATIVE AMERICAN (American Indian, Eskimo, Aleut)
4) WHITE
5) ASIAN OR PACIFIC ISLANDER
6) OTHER --Please specify:______________________

38. If you would like to make comments on health care delivery in your community, please do so. *(If more space is needed, continue on another page.)*

Thank you for your help.

Please return your questionnaire in the enclosed envelope to:

The Social and Economic Sciences Research Center
Washington State University
Pullman, Washington 99164-4014

Figure 7.4
A bad design for the back cover of a mail questionnaire

Between the Covers

Overall Ordering of the Questionnaire. In a well-designed survey, respondents are told in the cover letter what the study is about and why they should respond. The survey's theme is then echoed in the questionnaire's cover design. Imagine how people feel, then, if they turn to Page 1 and discover that the first questions have nothing to do with the main theme and are hard or embarrassing to answer. It is then, we think, that people are inclined to ask themselves whether they really want to participate in the survey.

In a study we once did on commercial fishermen, we wanted to include a series of difficult economic and technical questions. We were concerned that people would read the cover letter and still not see the relevance of the questions about boat sizes, engines, and the like. Eventually we decided to start the questionnaire by asking about fishing as a way of life. (For example, would they encourage their children to continue in their line of work? When did they decide to fish for a living?) These questions not only helped respondents get started but also caught their interest and gave them a context for later, more difficult questions.

The questions that go last are as important as those that go first. As a general rule, we suggest ordering questions on a scale from most interesting and topic-related at the beginning to those most likely to be objectionable at the end. Putting a question about people's income on Page 1 is like asking them to sign a blank check. Because they don't know yet how that personal information will be used, they are reluctant to provide it. However, if they have answered all the other questions, their objections may fade away.

Another rule we try to follow is to put questions on the same subject together. This eases the task of responding because people don't constantly have to switch from one topic to another in a seemingly arbitrary way. In a study that covers both housing and health issues, it would not make sense to ask first about the kind of housing structure in which one lives, then about visits to the doctor, then about rental payments, then again about choosing a medical clinic. Such a pattern would be disruptive and would make it hard for people to give reasoned answers.

When we design a questionnaire and have trouble ordering topics within it, we often ask for advice from potential respondents. For example, in a survey of employers, people told us it was easier for them to talk about what they expected to happen in their firms in the coming year *after* they answered questions about the preceding year.

Finally, within subject areas, we usually group questions that are similar in structure. For example, "yes" or "no" questions appear together, and so do those that ask respondents to rank a series of items.

The First Question. No questions get more scrutiny than the very first one or two. The beginning of the questionnaire is not the place for a question that could embarrass a respondent, that is hard to answer, is boring, or smacks of arcane qualities, which only a devoted questionnaire writer could love. Nor is it the place where one should put a question that does not apply to some respondents.

Therefore we advise the following:

1. Don't ask something open-ended: "How would you define a good leader?"
2. Don't ask something difficult: "Exactly how many times in the past year have you asked someone for medical advice?"
3. Don't ask something embarrassing: "Have you ever taken anything from a store without paying for it?"
4. Don't ask people directly about themselves: "Do you agree or disagree with the following statement—'I am a very cheerful, positive person.'" (If they are not cheerful and positive, you won't be off to a very good start with your questionnaire!)

Figure 7.5 shows how some questions work better than others at the beginning of a mail survey. The questions in Column A are likely to encourage respondents to continue with the survey. Questions A-1, A-2, and A-3 relate to the purpose of the study, seem useful and unbiased, and are easy to understand. In contrast, the questions in Column B are likely to discourage respondents. They seem unrelated to the survey (as in B-1), demand mental effort (B-2), and may give the impression that they are being asked if they agree with the researcher's point of view (B-3).

More Specifics About Question Order. Sometimes the order in which questions appear affects how people answer. Here's an example of what can happen:

Q1 When hiring new employees, should businesses in the community of Newton—15 miles away—give or not give first priority to people who live there?

1 GIVE FIRST PRIORITY

2 NOT GIVE FIRST PRIORITY

Q2 When hiring new employees, should businesses in our community give or not give first priority to people who live here?

1 GIVE FIRST PRIORITY

2 NOT GIVE FIRST PRIORITY

Research suggests that asking these types of questions in this order would result in fewer positive (GIVE FIRST PRIORITY) answers to Question 2 than if the order had been reversed. Question 1 tends to make people think generally about giving local residents priority for new jobs. Then, having already thought about the abstract principle of fairness, they apply it to their own community.

Figure 7.5

Good and bad examples of first questions on a mail questionnaire

Good (Column A)

For a survey about a new seat belt law:

A-1 Since January 1987, Washington drivers and front seat passengers may be fined for not wearing safety belts. Since that time, how often have you worn your safety belt? (Please circle the number of your answer.)

1 ALWAYS
2 MOST OF THE TIME
3 SOMETIMES
4 RARELY
5 NEVER
6 DON'T KNOW

For a survey of business owners:

A-2 In general, how would you describe the current business climate in your community compared to last year. (Circle the number of your answer.)

1 GETTING BETTER
2 ABOUT THE SAME
3 GETTING WORSE

For a survey about community needs:

A-3 How satisfied are you living in your present community?

1 VERY SATISFIED
2 SOMEWHAT SATISFIED
3 SOMEWHAT DISSATISFIED
4 VERY DISSATISFIED

Bad (Column B)

B-1 What is your birth date?

__ MONTH __ DAY __ YEAR

B-2 How do you feel about being asked to fill out and return mail questionnaires that ask you about your work or your company's business?

____
____
____

B-3 Please indicate whether you agree or disagree with the following statements.

My ideal community would be completely free of industrial air pollution.

1 STRONGLY DISAGREE
2 SOMEWHAT DISAGREE
3 NEITHER AGREE OR DISAGREE
4 SOMEWHAT AGREE
5 STRONGLY AGREE

In contrast, if people are asked about their own community *first*, they are likely to respond positively (GIVE FIRST PRIORITY). In other words, respondents are likely to say that local people in their own community should be given priority. Then, out of a sense of fairness, they feel they should reply positively to a subsequent question about another community as well.

Another example:

Q1 In general, how satisfied or dissatisfied are you with your marriage?

1 COMPLETELY SATISFIED
2 MOSTLY SATISFIED
3 NEITHER SATISFIED NOR DISSATISFIED
4 MOSTLY DISSATISFIED
5 COMPLETELY DISSATISFIED

Q2 In general, how satisfied or dissatisfied are you with your life?

1 COMPLETELY SATISFIED
2 MOSTLY SATISFIED
3 NEITHER SATISFIED NOR DISSATISFIED
4 MOSTLY DISSATISFIED
5 COMPLETELY DISSATISFIED

Question 1 gets people thinking about marriage as a central part of their lives, *before* thinking about their lives in general. The result is that people will tend to answer the second question differently than would otherwise have been the case. Had they been asked Question 2 first, they may have been more likely to weigh other aspects of their life more prominently, so that their answer was less affected by how they felt about their marriage.

It's hard to identify all the quirky effects that questions can have on each other. We suggest that you go through your questionnaire and try to separate pairs of questions in which the first might have a "carryover" effect on the second. Separation should help minimize the effect but may not eliminate it completely.

Transitions. When interviewers are about to change subjects in a face-to-face survey, they typically use some kind of transition. For example, they might say, "Now I'd like to ask . . . " or "That's all of the questions on where you spend vacations. Now I'd like to ask a few questions about the kind of books you may have read last year."

Transitions are natural and helpful, like lane change and exit signs along interstate highways. They also add a conversational tone, something easy to forget in mail surveys. Without transitions, questions begin sounding like a litany of demands: "How much? Why? When? How often? Where?" We suggest transitions

- *regularly*, when starting a new line of inquiry; for example, "Next we would like to know your opinions about how to increase seat belt use";
- *sometimes*, at the top of the first of two facing pages, to catch the eye of a respondent who flips through the questionnaire looking for clues as to what it contains; and
- *occasionally*, to break the monotony of long series and to provide motivation.

Transitions should not provide a complicated, analytic justification for questions that follow but should convey the message that the questions are important. Some transitions signal major changes that need justification, and others indicate only a slight change in emphasis.

The transition to personal questions at the end of the survey should be added almost parenthetically, as if the information were less important. You can make personal questions seem less threatening with a statement like "Finally, we'd like to ask you a little about yourself."

Individual Page Design. We now turn to the final step—taking the ordered questions and fitting them on individual pages. An unfortunate myth often guides this process. The myth is that fewer pages are always better, even if it means reducing the type size and crowding questions into nooks and crannies on the smallest amount of paper possible. Hence, some questionnaires consist of oblong blocks of questions of all different sizes packed together. Respondents end up weaving back and forth, across, and up and down every page. In a misguided attempt to remedy the situation, some surveyors demarcate questions with thick black lines, making the page look even more cluttered. And sometimes they compound the problem by adding prominent coding and keypunching instructions. In the end, the least dominant features of the page are the questions or answer choices. (See Figure 7.6.)

We start with a very simple premise—the questionnaire is *for the respondent*. Therefore, the ink should be used for questions and answer choices that the respondent needs to see. Our goal is to make it as easy as possible for people to go from one question to the next

SECTION 2 – TEACHING EXPERIENCE – *Continued*

NOTE: *Answer item 7 ONLY if you marked box 6 in answer to question 6 on page 4.*

7. Which of these categories best describes your previous position in the field of education?

Mark (X) only one box.

019
1 ☐ Administrator (e.g., principal, assistant principal, director)
2 ☐ Counselor
3 ☐ Librarian/media specialist
4 ☐ Coach
5 ☐ Other professional staff (e.g., department head, curriculum coordinator)
6 ☐ Instructional aide
7 ☐ Noninstructional support staff (e.g., secretary)

Skip to item 9

NOTE: *Answer items 8a–e ONLY if you marked box 7 in answer to question 6 on page 4.*

8a. For whom did you work? – *Record the name of the company, business, or organization.*

b. What kind of business or industry was this? *For example, retail shoe store, State Labor Department, bicycle manufacturer, farm.*

020

c. What kind of work were you doing? *Please record your job title; for example, electrical engineer, cashier, typist, farmer, loan officer.*

021

Figure 7.6
Part of a page from a poorly designed mail questionnaire

without inadvertently skipping one or being confused as to what to do next. Figure 7.7 illustrates how a questionnaire page looks when respondents are top priority. Here are some of the most important features.

1. The black lines often used to separate questions are completely eliminated in favor of an open format.
2. A vertical flow is established. In other words, respondents generally move vertically down the page, reading first the question and then the answer choices lined up one below another.
3. The questions are in dark (or boldface) type and the answers in light. Respondents are often unaware of this difference, yet it is a feature that guides them through the questionnaire. Another way to accomplish the same effect is to use lowercase letters for questions and capitals for answers.
4. Brief instructions on how to answer the questions are given exactly where the information will be used, not at the beginning of the questionnaire or even on a preceding page. Wherever possible, graphic design is used to help communicate the instructions. For a series of questions in which the procedure is the same, the instructions can be given once—"Circle the number of your answer"—then abbreviated—"Circle number"—and eventually omitted.
5. Numbers are used for answer categories rather than boxes or fill-in-the-blank lines. This practice "precodes" the answers so they can be entered directly into a computer. Unless you are

Figure 7.7

Observing the key to effective layout in mail questionnaires

Q23 **What is your present marital status?** (Circle the number of your answer.)

1 NEVER MARRIED
2 MARRIED
3 DIVORCED
4 SEPARATED
5 WIDOWED

Q24 **What is the highest level of education you have completed?** (Circle number.)

1 NO FORMAL EDUCATION
2 SOME GRADE SCHOOL
3 COMPLETED GRADE SCHOOL
4 SOME HIGH SCHOOL
5 COMPLETED HIGH SCHOOL
6 SOME COLLEGE
7 COMPLETED COLLEGE
8 SOME GRADUATE WORK
9 A GRADUATE DEGREE

Q25 **What is the highest level of education your parents have completed?**
(Circle number of one choice in each column.)

MOTHER	FATHER	
1	1	NO FORMAL EDUCATION
2	2	SOME GRADE SCHOOL
3	3	COMPLETED GRADE SCHOOL
4	4	SOME HIGH SCHOOL
5	5	COMPLETED HIGH SCHOOL
6	6	SOME COLLEGE
7	7	COMPLETED COLLEGE
8	8	SOME GRADUATE WORK
9	9	A GRADUATE DEGREE

Q26 **Do you own or rent the home in which you now live?**

1 OWN HOME → IF YOU OWN YOUR OWN HOME, SKIP TO Q-30, ON THE NEXT PAGE
2 RENT HOME
↓

Q27 **(If you rent) How much is your monthly rent?**

1 ZERO
2 MORE THAN ZERO AND LESS THAN $200
3 $200 TO $399
4 $400 TO $599
5 $600 OR MORE

going to tabulate the results by hand, precoding saves lots of time. (See Chapter 10 on data entry.)

6. Answer choices are written in such a way that at least one applies to every respondent. Otherwise, answers are ambiguous. For example, a NO response to the question "If you are over 50, are you going to retire within 10 years?" might mean "I'm not over 50" *or* "I'm not going to retire within 10 years."
7. Use a multiple-column design when two or more questions (with the same answer categories) can be combined into one. This format looks attractive and conserves space.
8. For questions that don't apply to everyone—the "skip" questions—arrows are used to show people where to go next. An arrow out to the right tells people to skip to a later question; an arrow pointing down leads to the next question (see Q-26 of Figure 7.7). Just to double-check, the next question starts out with "If you rent" (or whatever is appropriate).
9. The answer ZERO is clearly distinguishable and won't be confused with NO RESPONSE. "Zero" rent and not choosing any of the dollar categories are two very different responses that can cause serious headaches in data analysis unless clearly differentiated.
10. Questions are made to fit each page so that respondents will not overlook anything. If items in a series must be split between two pages, the headings are repeated on the next page.

Series of questions for which people are asked to use the same categories pose special design problems. Sometimes surveyors use a row/column format that requires people to match each row and column and then put some mark in the blank space. Sometimes they are even required to rank each item in the same process (see Figure 7.8). Don't do it! Always have something written at the intersection of each row and column (as shown in Figure 7.9); doing so makes the respondent's job much easier. And if ranking is involved, make it a separate question. In Figure 7.9 the questionnaire has an even-looking, block format, and dotted lines (called **leaders**) connect questions with answer choices.

When asked for advice on designing a questionnaire, we sometimes feel like the real estate agent describing the three most important determinants of home value—location, location, and location. For a questionnaire, it is consistency, consistency, and consistency. Whatever format you choose, use it consistently.

Figure 7.8
An example of items in a series that are hard to answer because of poor formatting

Q1 Listed below are some ideas suggested as possible goals for developing our local economy. Please indicate whether you feel that each goal should NOT be a priority, should be given a LOW priority, MEDIUM priority, or HIGH priority.

	Not a priority	A low priority	A medium priority	A high priority	Rank three most important here
Increase the number of job opportunities by encouraging the development of our existing business districts					
Encourage the development of new industries in the area					
Encourage the development of cottage or home-based industries in the area					
Protect our natural environment from activities that are damaging to it					
Involve more citizens in planning for the community's economic future					
Preserve the present agricultural land in the Wahkiakum/Naselle area					
Preserve and maintain our historic assets					
Preserve the present way of life in our communities					
Encourage the development of tourism in the area					
Encourage more local government support for and involvement in economic development					

Are there any others?_______________

Figure 7.9

An example of a good format for items in a series

Q-1 Listed below are some ideas suggested as possible goals for developing our local economy. Please indicate whether you feel that each goal should NOT be a priority, should be given a LOW priority, MEDIUM priority, or HIGH priority.

Goal number	Possible goals	**How much priority, if any, should each goal have? (Please circle your answer.)**			
1	Increase the number of job opportunities by encouraging the development of our existing business districts..................	NOT	LOW	MEDIUM	HIGH
2	Encourage the development of new industries in the area...............................	NOT	LOW	MEDIUM	HIGH
3	Encourage the development of cottage or home-based industries in the area.........	NOT	LOW	MEDIUM	HIGH
4	Protect our natural environment from activities that are damaging to it............	NOT	LOW	MEDIUM	HIGH
5	Involve more citizens in planning for the community's economic future...........	NOT	LOW	MEDIUM	HIGH
6	Preserve the present agricultural land in the Wahkiakum/Naselle area.............	NOT	LOW	MEDIUM	HIGH
7	Preserve and maintain our historic assets.....	NOT	LOW	MEDIUM	HIGH
8	Preserve the present way of life in our communities..........................	NOT	LOW	MEDIUM	HIGH
9	Encourage the development of tourism in the area...............................	NOT	LOW	MEDIUM	HIGH
10	Encourage more local government support for and involvement in economic development.....................	NOT	LOW	MEDIUM	HIGH
	Are there any others? (Please list.)				
11	______________________	NOT	LOW	MEDIUM	HIGH
12	______________________	NOT	LOW	MEDIUM	HIGH

Q-2 Of the possible goals listed in Q-1, which do you feel are most important for our area? (Please write the goal number from Q-1 in the appropriate box.)

☐ Most important ☐ Second most important ☐ Third most important

Highway engineers would not ask drivers to move back and forth, from side to side on the interstate, but we have seen many questionnaires in which the style of question randomly shifts. This practice makes the respondent's task much more difficult than necessary.

Remember that white space is a virtue. Using unnecessary dividing lines is like putting occasional stop signs, or at the very least, unnecessary dividers in the way of the questionnaire flow.

Putting It All Together

If you follow our earlier recommendation to make your questionnaire with legal sheets folded in half, you will quickly discover that the survey booklet consists of four-page sets. You cannot just add a single page when all your questions don't fit, you have to add four.

Although we generally prefer fewer sheets of paper, we cannot always condense our spacing to make things fit. The alternative is to add the extra pages and space things out. In the process, we look carefully at our paging so that respondents don't have to turn a page in the middle of an item, reading the question on one side and the answer choices on the next. Sometimes we rearrange the order of questions somewhat to better fill each page. This practice avoids having blank space on some pages and others that are crammed full.

The final result should be a questionnaire that looks professional and attractive and, most of all, encourages people to respond.

Pretesting

Trying out the questionnaire before an actual survey is like test-driving a new car. The purpose is to learn whether it works to your satisfaction or has big problems. Not only must you check all the individual parts, but the total effect must be evaluated as well. "Test-driving" a questionnaire is time-consuming but absolutely essential.

Pretesting is far more than sending questionnaires to a sample of respondents and then counting how many come back. If you do that, you will miss important information about how the questionnaire works. For example:

- Is each question getting the information it is intended to get?
- Are all the words understood?
- Are the questions interpreted the same by all respondents?
- Do all close-ended questions have an answer that applies to each respondent?
- Does the questionnaire create a positive impression that motivates people to respond?

- Are the questions answered correctly and in a way that can be understood?
- Are skip patterns followed correctly?
- Does any part of the questionnaire suggest bias on your part?

Pretesting is best done in two phases. The first is to talk with the people who will use the survey results—if you are preparing the survey for someone else. The information users might be, for example, policymakers or agency administrators. Because they have substantive, practical knowledge about the kind of data that are being collected, they can spot technical problems that surveyors might miss. We once worded a question that suggested police officers were responsible for fining as well as arresting traffic violators. It was quickly pointed out to us that judges fine traffic violators, and we were saved from considerable embarrassment.

The second phase of the pretest should be conducted with people who are typical of likely respondents. It is important that the diversity of the population be represented among those who pretest the questionnaire, especially with respect to characteristics that are expected to affect the way people answer. For example, to pretest a survey about the credit needs of local business owners, the questionnaire should be given to owners of large and small businesses, men and women, and so forth.

It is important to watch people fill out questionnaires in person rather than simply mailing them a form. This way, we can watch for signs that people are puzzled, that they have misread instructions and have to change answers—or other things that indicate problems. Also we try to debrief people as soon as they complete the questionnaire. This helps us identify problems that weren't obvious from gestures and facial expressions.

Telephone Questionnaires

Writing for the eye—for print, for reading—is different than writing for the ear. In the 1950s, when television began to capture the attention of millions of viewers in the U.S., newspaper journalists who went to work for TV stations had to come to grips with what broadcasting meant for their craft. Writing for the ear entailed an entirely different pace, a different style, and even a different delivery.

When we design mail questionnaires, we write for the eye. When we shift to the telephone, we write for the ear. Telephone questionnaires depend entirely on verbal communication and must *sound* rather than *look* professional. We don't have to motivate respondents with our design, but it is essential that we use layout to make the interviewer's job as easy as possible. Features we would never use in a mail questionnaire are common in one used for telephone interviews; to name just a few: extra instructions on how to administer the interview, data entry information, space for coding and writing down comments, and page turns that do not separate questions that need to appear together.

Given all these differences, it isn't surprising that even the best mail questionnaires don't work on the telephone. To convince yourself of this, try reading the questions in Figure 7.7 out loud to a friend and asking him or her to respond. You will probably run into problems right away. It is unclear which words you should read to your friend and which are intended for you only (for example, the instruction "If you own your own home, skip to Q30"). Answer choices that are listed separately in the mail version must now be incorporated into the question stem, as in the following.

Q-23 Which of the following best describes your present marital status: never married, married, divorced, separated, or widowed?

NEVER MARRIED.................. 1
MARRIED.......................... 2
DIVORCED......................... 3
SEPARATED........................ 4
WIDOWED.......................... 5
REFUSAL/NO ANSWER........... 99

Finally, ideas that are conveyed graphically to people reading the survey now need to be clarified as part of the oral question. The notion of using one column for mother's and one for father's education (as in Q25 of Figure 7.7) means little when read out loud.

The best check for a telephone questionnaire is whether it can be read to most respondents exactly as written. The interviewer should be able to clearly communicate all of the questions without having to add spontaneous interpretations or additional instructions.

In a telephone survey, the person being interviewed can and often does ask for more information. Sometimes they want to know more about specific questions, for example, "Should I count stopping by to pick up a prescription from the receptionist as a visit to the doctor?" Sometimes they want to know why a certain question is being asked, or something more general, such as the group or organization sponsoring the study. Interviewers must have ready answers to these questions, either on the survey form itself or on help sheets (which we discuss in Chapter 8).

We assume here that the interviewer will be writing down answers on a questionnaire rather than using a computer-assisted method. Computer-assisted telephone interviewing (CATI) systems are efficient but also expensive (we describe them in Chapter 8). Hence, for people who want to do their own survey the cost and effort required for a CATI system will likely outweigh the benefit. And it is important to note that the quality of data one obtains from each method is not inherently different.

Wording the Questions

Sit down in a room with a draft of your questionnaire and someone who can act as the respondent. Sit so that neither of you can see the other, and then start the interview. Quickly, the fundamental difficulty associated with telephone interviewing will become clear. You must read each word to the respondent, one at a time. In turn, he or she must remember each word, one at a time, and so build the meaning of the question. This will likely require considerable concentration and effort for both of you. A couple of tries will help you discover for yourself one of the most important principles for writing telephone survey questions—*keep them short and simple.*

Sometimes, questions have a long series of answer choices, as in "Are you very satisfied, quite satisfied, somewhat satisfied, slightly satisfied, neither satisfied nor dissatisfied, slightly dissatisfied, somewhat dissatisfied, quite dissatisfied, or very dissatisfied?" Over the telephone this question is comical, even though it can work when someone reads a properly formatted written version to themselves. This is not to say that all questions must be limited to "Are you generally satisfied or dissatisfied?" However, as this example shows, questions with many answer choices can be very awkward when read out loud.

When it is necessary to have many answer choices, we suggest using a two-step approach. For example, ask first whether the person is generally satisfied or dissatisfied. Then, if they pick *satisfied,* ask whether they are very, somewhat, or only slightly satisfied.

Asking people to rank items is difficult over the phone. The first example in Figure 7.10 shows how hard it is for people to remember numerous choices and then pick the one they like best. Even when people remember all the choices, they tend to pick from among the last-mentioned categories rather than the first. We suggest taking respondents through a process in which they are asked to rate each item individually and then, at the end of the question, to choose the most important. The second example in Figure 7.10 shows how this is done.

Figure 7.10

Different ways of asking respondents to rank items in a telephone interview

Difficult for respondents to rank in a telephone interview:

People have suggested many reasons why the National Park Service should reintroduce wolves into Yellowstone National Park. I am going to read you a list of six reasons that have been mentioned. Please tell me which you believe is most important, second most important, and third most important.

The park's ecosystem would be more complete if wolves were reintroduced. . . . . ☐

Visitors would have an opportunity to hear wolves howl . ☐

People are responsible for the wolves' demise and therefore have a responsibility to reintroduce the species . . . . ☐

Visitors should have an opportunity to see wolves in their natural habitat . . . . ☐

A natural predator like the wolf would help keep elk and deer populations at a level that the park could support . . . . ☐

Yellowstone Park is an important test case that would teach us how to reintroduce species in controlled situations . . . . ☐

Easier for respondents to rank in a telephone interview:

People have suggested many reasons why the National Park Service should reintroduce wolves into Yellowstone National Park. I am going to read you a list of six of those that have been suggested. For each, please indicate whether you think it is not important, somewhat important, or very important.

Continued on next page

Less-experienced interviewers often improvise lead-ins and connecting words for questions with many answer choices. For example, in the first wolf question in Figure 7.10, they might say, "The park's ecosystem would be more complete if wolves were reintroduced *versus* visitors would have an opportunity to hear wolves howl *compared to* . . . " Improvising in this way is a very bad practice because different interviewers end up reading the questions in different ways. Hence, the answers people give are not really comparable.

Figure 7.10
continued

Goal	Possible goals	Not important ∇	Somewhat important ∇	Very important ∇
A	The park's ecosystem would be more complete if wolves were reintroduced......	1	2	3
B	Visitors should have an opportunity to hear wolves howl............................	1	2	3
C	People are responsible for the wolves' demise and therefore have a responsibility to reintroduce the species.................	1	2	3
D	Visitors should have an opportunity to see wolves in their natural habitat............	1	2	3
E	A natural predator like the wolf would help keep elk and deer populations at a level that the park can support	1	2	3
F	Yellowstone Park is an important test case that would teach us how to reintroduce speces in controlled situations............	1	2	3

Now, of the reasons we have just discussed, which one do you think is the most important? If you like, I would be happy to read the list again. (INTERVIEWER: PUT GOAL LETTER IN BOX.)

☐ Most important

Which one is the second most important?

☐ Second most important

Which one is the third most important?

☐ Third most important

To control how answer choices are read over the phone, a good practice is to include them in the body, or stem, of the questions. The second wolf question in Figure 7.10 is an example: "For each [reason], please indicate whether you think it is not important, somewhat important, or very important." Besides preventing interviewers from improvising, this format also helps maintain a conversational tone in the interview.

Finally, in telephone questionnaires, we often include additional answer choices that we wouldn't include in a mail version. They are "Don't know" and "Refusal." We don't read these as part of the question because we would rather respondents didn't choose them. However, sometimes they are appropriate, and we need a way to record them.

The Interviewer's Introduction

Imagine walking in the door after work, answering the phone, and hearing the following:

> Hello, my name is Sam and I'm calling from New Jersey. We're doing a study on drinking, drug use, and mental health. Even if you don't use drugs or alcohol, we'd still like to talk to you. Can you please tell me how many people live in your household?

Chances are, your reaction to this telephone survey will not be especially positive. The interviewer failed to completely identify himself or the survey. He then raised a red flag right away by saying that you

Figure 7.11
How to word the introduction in a telephone interview when the respondent's name has been selected from a directory or list

Q1 This is ____________ calling from Washington State University. May I please speak with ____________ ?

Hello, my name is ____________, and I'm calling from Washington State University in Pullman. We have been asked by the Washington State Department of Ecology to talk with people about their opinions on environmental issues affecting people who live in this state.

. . . .

INSERT QUESTION ON ADVANCE LETTER IF APPLICABLE.*

. . . .

This interview is completely voluntary and confidential. The survey will take about 15 or 20 minutes. If I come to any question that you would prefer not to answer, just let me know and I'll skip over it. OK?

1 YES
2 NO ⟶ [INTERVIEWER READ: When would be a better time to call you back? INTERVIEWER WRITE: TIME AND NAME ON CALL RECORD.]

*Have you received the letter that we sent you about this study?
1 YES
2 NO [It was a short letter explaining the purpose of the survey.]
3 DON'T KNOW [It was a short letter explaining the purpose of the survey.]

are important *even if you don't use drugs or alcohol,* a less-than-respectful introduction. And finally, he launched into his questions immediately, without giving you time to register what was happening.

We take a different approach to beginning telephone surveys because the first few minutes of the interview are so critical to the success of the whole survey. People typically decide whether to respond based on what they hear right after picking up the phone. Therefore, we try to set a neutral, businesslike tone and moderate pace right from the start. Our standard format for introductions includes the following:

- the interviewer's name;
- the organization and *city* from which he or she is calling;
- a one-sentence description of the survey; and
- a conservative estimate of how long the interview will take.

This information tells prospective respondents who is calling and why. It also helps give the interviewers credibility and differentiates them from callers trying to make a sale or solicit contributions. Before ending the introduction, we explain that the interview is voluntary and confidential. We then informally ask the respondent for permission to begin by saying, "If I come to any question that you prefer not to answer, just let me know and I'll skip over it. OK?" This technique slows the pace and helps to begin the interview on a relaxed note.

The preface to this introduction depends on the sampling method being used (see Chapter 5). If respondents have been selected from a list, interviewers simply ask for the appropriate person and begin their introduction as soon as that person is on the line (Figure 7.11). If phone numbers have been selected with a random or add-a-digit technique, interviewers should verify that they have reached the correct number and then use the appropriate within-household (or within-business) selection technique (Figure 7.12).

Ordering the Questions

We start a telephone questionnaire in essentially the same way as a mail questionnaire. The beginning is not the place for a question that has numerous response choices, concerns an unexpected topic, requires a long answer, or is confusing in any way.

Be sympathetic to your respondents. It is very difficult to grasp ideas delivered one word at a time without any visual clues. Some people will think that they have not much to say and would really rather you pick someone else. Most are unaccustomed to being interviewed over the telephone and will say to themselves as you begin,

Figure 7.12

How to word the introduction in a telephone interview when random digit or add-a-digit dialing has been used and the respondent must be selected from within the household

Q1 Hello, is this ____________? [IF NO, INTERVIEWER READ: "I'm sorry, I have the wrong number." IF YES, CONTINUE.]

Q2 Is this a residential telephone? [IF NO, INTERVIEWER READ: "I'm sorry, I have the wrong number." IF YES, CONTINUE.]

Q3 My name is ____________, and I'm calling from Washington State University in Pullman, Washington. We have been asked by the Washington State Department of Ecology to talk with people in Washington about their opinions on environmental issues affecting people who live in this state. We need to be sure we give every adult a chance to be interviewed for this study. Thinking only of adults in your household—that is, people 18 years of age or older, which one had the most recent birthday, you or someone else?

1 SELF [INTERVIEWER: Skip to Q-5.]
2 SOMEONE ELSE [INTERVIEWER: Ask to speak to that person. If that person is not home, ask when might be a good time to reach him or her.]

Q4 My name is____________, and I'm calling from Washington State University in Pullman, Washington. We have been asked by the Washington State Department of Ecology to talk with people in Washington about their opinions on environmental issues affecting people who live in this state.

Q5 This interview is completely voluntary and confidential. The survey will take about 15 or 20 minutes. If I come to any question that you would prefer not to answer, just let me know and I'll skip over it. OK?

1 YES
2 NO ⟶ [INTERVIEWER READ AS APPROPRIATE: When would be a better time to call you back? May I have a name so I know who to ask for? INTERVIEWER WRITE: TIME AND NAME ON CALL RECORD.]

"So this is how a telephone interview works . . . " Some who agree to your interview will be anxious and will still be deciding whether they really want to proceed.

This is why we start with questions that are topically related to what the respondent already knows the interview is about and then move toward subjects that may seem less related. We also use question order to minimize the new ideas and instructions that have to be understood. We group questions on the same topic, as well as similarly structured questions within topic areas. We order subjects so that they seem logical to respondents. Finally, we always try to avoid situations in which respondents have to repeat information they provided earlier. For example, we wouldn't ask a farmer about production plans for next year and then ask how business was this year.

The First Questions

As a general rule, the very first question should be close-ended with no more than two or three answer choices. A yes/no format works well. The second or third question should be open-ended. That allows respondents to become comfortable on the phone—to find their answering voice—and it helps pace the interview.

Why is pace important? When inexperienced interviewers and respondents are nervous, the interview can take off like a greyhound. Questions are read too quickly, and respondents, in an effort to be cooperative, feel compelled to answer in kind. Starting with one or two close-ended questions helps show respondents how the interview works. Then, an open-ended question slows the pace and helps set a natural and conversational tone.

Good opening questions in a survey of farmers might be

Q1 I would like to begin by asking whether or not you were raised on a farm.

1 YES

2 NO

Q2 What are some of the reasons why you now live on a farm?

Sometimes people insert the most critical questions at the start of the interview because they think respondents might hang up before getting to the point of the survey. In a word: Don't! It is more important to get people past the first few minutes, when they may not be concentrating as hard on your questions. Experienced athletes never try to perform at full speed without first warming up. A similar perspective on what we expect of respondents is helpful.

Transitions

Imagine you are riding in a car and telling the driver how to reach a particular destination. It is much better for you to give ample warning—"At the next corner we are going to turn left"—than to wait until an intersection and say "Turn now!" Similarly, it is helpful to respondents if you provide ample warnings—"Next I'd like to ask you several questions about your current job." Transitions like these give people a sense of where the interview is headed and make it easier to understand the questions when they are asked (Figure 7.13).

Adding comments to signal the end of one subject and the beginning of another is also helpful; for example, "That's all of the questions about where you live, and now I'd like to ask a few questions about any additional education you might have had since completing high school."

We add transitions only after we are satisfied with how the questions are ordered. They improve the tone and pace of the telephone interview. They also help build rapport and emulate what one would do naturally in a conversation. Finally, they help avoid awkward silences by filling in while the interviewer is writing down answers and turning pages.

Designing the Pages

Well-designed mail and telephone questionnaires look very different because they are not intended for the same audience. In a mail questionnaire, we always have the respondent in mind and try to create a format in which questions *look* easy and *are* easy to answer. In contrast, respondents never see telephone questionnaires, but interviewers do, again and again in one interview after another. While cradling a telephone receiver (unless headsets are available), they have to turn pages without allowing long pauses in the dialogue, write down answers that are all too often delivered quickly, and, most important record responses accurately.

Hence, good graphical design and consistency are important in telephone questionnaires because they help make the interviewer's job routine. The goal is to let interviewers concentrate on listening to respondents, probing for more information when necessary, and

Figure 7.13
How and why to use lead-ins, transitions, and fillers in telephone questionnaires

An example . . .	*And why . . .*
Now I would like to ask you a few questions about an ongoing project at the Hanford Nuclear Reservation.	Signals change in subject.
It is called the Environmental Dose Reconstruction Project. This is an additional subject we are gathering data on.	Slows the pace.
The purpose of the project is to determine what level of radiation people in the state may have been exposed to from 1944 to 1970 due to Hanford activities.	Provides background information to build interest.

recording accurate information. They should never have to improvise because they are uncertain about what to do next.

Figure 7.14 illustrates the main principles of page design in telephone questionnaires. They are:

1. Select one type style and use it consistently for every single word and phrase that is *always* read. We suggest using **bold type** if the questionnaire is written on a word processor; otherwise, use lowercase letters. The mental clue this sends to interviewers is that, if a word or phrase is **bold,** it is to be read without fail.
2. Select another type style (*italics* perhaps) for words or phrases that are sometimes read to respondents, for example, a probe such as "*If you aren't sure, you can give me your best guess.*" Definitions or explanations that apply to a specific question are also done in italics and are placed right after the question.
3. Use a third type style (capital letters, perhaps, but not bold) for special instructions to the interviewer, which are never read to the respondent.
4. Place answer choices toward the right side of the page (rather than on the left as we did for mail) with the code numbers to the right of the answers. This format is the easiest for interviewers, assuming they are right-handed.
5. Include in your list of answers the responses that will not be read out loud but may be needed. Examples are NO OPINION, DOES NOT APPLY, and REFUSED. Also included are answer categories already read to respondents as part of the question. Give the right signal to the interviewer by using the type style reserved for answers that are never read out loud.
6. Be consistent about using certain numbers for specific purposes. For example, you might reserve 7 or 77 for NO OPINION, 8 or 88 for DOES NOT APPLY, and 9 or 99 for REFUSED. Assign lower numbers to negative responses (1 for NO or DISAGREE) and higher numbers to positive responses (2 for YES or AGREE). Being inconsistent in your use of numbers will almost certainly lead to interviewer errors. Not surprisingly, interviewers become accustomed to a particular convention and then have trouble adjusting to subtle changes.
7. Use a consistent, clear format to explain and *illustrate* skip patterns. Begin questions that apply to all respondents at a standard, left-hand margin. If a respondent who answers "yes" should continue and one who answers "no" should skip, then (a) draw an arrow from the "yes" answer to the next question

Figure 7.14

Observing the keys to effective page design in telephone questionnaires

Q23 **Which of the following best describes your present marital status: never married, married, divorced, separated, or widowed?**

NEVER MARRIED..... 1
MARRIED.............. 2
DIVORCED............ 3
SEPARATED........... 4
WIDOWED............ 5
REFUSED/NO ANSWER 99

Q24 **What is the highest grade you completed in school?**
[CODE ANSWER FROM THE CATEGORIES BELOW.]

HIGH SCHOOL OR LESS...................... 1
SOME COLLEGE/BUSINESS SCHOOL/
OR VOCATIONAL-TECHNICAL TRAINING 2
BACHELOR'S DEGREE....................... 3
MASTER'S OR DOCTORATE DEGREE........ 4
REFUSED/NO ANSWER...................... 99

Q25 **What is the highest grade your parents completed in school?**
(If you aren't sure, you can give me your best guess.)

[CODE ANSWER FROM THE CATEGORIES BELOW.]

	MOTHER	FATHER
HIGH SCHOOL OR LESS......................	1	1
SOME COLLEGE/BUSINESS SCHOOL/ OR VOCATIONAL-TECHNICAL TRAINING	2	2
BACHELOR'S DEGREE.......................	3	3
MASTER'S OR DOCTORATE DEGREE........	4	4
REFUSED/NO ANSWER......................	99	99

Q26 **Do you own or rent the home in which you now live?**

OWN HOME........... 1 → SKIP TO Q-30
RENT HOME.......... 2
REFUSED/NO ANSWER 99

Q27 [If R RENTS] **How much is your monthly rent?** [CODE ANSWER
INTO CATEGORIES BELOW.]

ZERO...................................... 1
MORE THAN ZERO AND LESS THAN $200 2
$200 TO $399............................ 3
$400 TO $599............................ 4
$600 OR MORE........................... 5
DOES NOT APPLY......................... 88
REFUSED/NO ANSWER.................... 99

and (b) draw an arrow from the "no" answer either to the appropriate question (if it is on the same page) or to a box with skip instructions (if it is not). As a general rule, indent by about five spaces questions that some people are instructed to skip.

8. Make sure interviewers don't have to turn pages in the middle of a question or during a complicated skip sequence. This means that telephone questionnaires often have some pages that are only partly filled.

Printing the Questionnaire

Convenience of the interviewer remains the major objective. If questionnaires are long, they can be printed as booklets. Shorter questionnaires can be stapled in the upper left-hand corner.

We sometimes use colored paper for the cover pages of surveys in which various subgroups of respondents get different versions of the questionnaire. For example, we might use yellow for small businesses and blue for large. When we have a long questionnaire in which large sections are skipped, we identify the different sections with various light colors of paper. Again, the purpose is to help interviewers work as efficiently as possible.

Pretesting

Taking a telephone questionnaire for a test drive is no less important than it is for a mail survey. The purpose is the same—to make sure the questionnaire measures what it is supposed to, that people understand and can easily answer the questions, and that interviewers can work with it.

Sometimes, the results are distressing because people typically concentrate on how the questionnaire looks, reading it to themselves as they design and edit. Inevitably, problems arise as soon as questions are read to someone who is not within sight. We therefore strongly recommend pretesting over the telephone. To work out the major bugs, we do the first pretest interviews with people who worked on the survey (and who may be surprised at how it sounds when read out loud). Then, to identify additional problems, we draw a small sample of actual respondents and interview them just as we would in the real study.

As we have emphasized throughout this section, the interviewer is the telephone questionnaire's major user. Hence, it is important that several different interviewers be involved in the pretest. One or two may miss certain problems (especially if they helped write the questionnaire), so we suggest getting a number of interviewers together for the trial run. This helps ensure that you will identify any awkward, unclear skip instructions or other rough spots that interfere with getting quality data.

One of the most useful techniques we have found for pretesting questionnaires is to bring together some of the interviewers who will be doing the survey. We have each conduct as many interviews as they can during a particular period of time. Then, we debrief them as a group, going through the questionnaire item by item.

Face-to-Face Questionnaires

In most respects, face-to-face questionnaires are similar to those used in telephone surveys. Most important, they are written so that people can respond based primarily on what they hear. Rather than repeat what has already been said about telephone questionnaires, we will only point out a few ways in which face-to-face questionnaires are different. First, interviewers have personal contact with respondents—they can watch how respondents react to questions, see when people hesitate, and then clarify or explain as they need to. Second, they can use visual aids to make sensitive questions less threatening or complicated questions more simple. For example, when asking for income information, they can use flash cards with preprinted categories (see Figure 7.15). Finally, face-to-face questionnaires can include very complex questions that would be difficult for respondents to follow over the phone. Again, using visual aids is usually possible.

Often, respondents like to follow along with the interviewer on a blank copy of the questionnaire. We use a special version without interviewer or coding instructions. Sometimes we put see-through plastic covers on the pages so people will not mark their answers on the questionnaire. We have also seen face-to-face surveys in which people actually fill out a section of the questionnaire on their own, for example, when they are asked a long series of attitude questions that

Figure 7.15
Example of a flashcard visual aid

**Total Family Income in 1993–Before Taxes,
All Members of Family Living in your Household**

A.	Under $10,000 a year	(or under $192 a week)
B.	$10,000 to $24,999 a year	(or under $192 to $480 a week)
C.	$25,000 to $49,999 a year	(or $481 to $961 a week)
D.	$50,000 to $74,999 a year	(or $962 to $1,442 a week)
E.	$75,000 and over a year	(or $1,443 or more a week)

they can answer faster by reading to themselves. Such flexibility is fine, as long as the questionnaires are filled out accurately and legibly.

We suggest putting an attractive cover on face-to-face questionnaires, much the same as we do in mail surveys. We do this for the respondents, who will see the questionnaire and may be influenced by its outside appearance.

In summary, if you are doing a face-to-face survey, we recommend that you carefully read the section in this chapter on telephone questionnaires. Remember that your main objective still is to help the interviewers do the best and most efficient job possible.

Summary

Designing a questionnaire is like constructing a building: *Questions* are to the first as *support columns* are to the second. Both are absolutely critical to success and, for the most part, independently built. How they are placed relative to each other and then connected are matters that should never be left to chance.

Designing a questionnaire means ordering questions, writing transitions, and arranging the results, one page at a time. Many principles guide the design of a successful mail questionnaire, while others apply to forms used in telephone and face-to-face surveys.

It is critical to recognize that self-administered questionnaires are written for people to read, whereas interview questionnaires are written for them to hear. In mail surveys, our goal is to design a visually appealing questionnaire that encourages people to respond, thereby reducing the potential for nonresponse error. In interview surveys, our goal is a questionnaire that communicates effectively to the interviewer and sounds good when read to the respondent. Again, the result is a higher response rate. In both cases, a well-designed questionnaire helps ensure that people's answers are as accurate as possible.

Having a final questionnaire in hand is reason to celebrate. Progress is being made! Still, the task of asking and encouraging people in the sample to complete the questionnaire is just beginning. It requires that a whole new set of concerns be addressed, which we do in Chapter 8.

For more details on questionnaire design, see the following:

Chapter 4, "Constructing Mail Questionnaires," and Chapter 6, "Constructing Telephone Questionnaires," in *Mail and Telephone Surveys: The Total Design Method*, by Don A. Dillman. Wiley- Interscience, New York, 1978.

Chapter 4, "Issues in Questionnaire Design and Writing," in *Survey Research by Telephone*, by James H. Frey. Sage Publications, Newbury Park, CA, 1989.

8

Setting Your Survey in Motion and Getting It Done

In this chapter, we explain how to carry out mail, telephone, face-to-face, and drop-off surveys. Each of these methods is different in terms of the resources needed and the schedule. No matter which method seems right, it is important to read this chapter before committing to one or another. *Make sure* you can put together the necessary combination of supervision, office help, equipment, and supplies—all coordinated in the required amount of time.

Our main concern in this chapter is how to increase response rates so as to avoid nonresponse error. Hence, the step-by-step procedures presented here are designed to produce acceptably high response rates. This goal translates into different procedures for each method, but the reasoning is always the same: *People are more likely to respond when they think the benefits outweigh the costs, when they think they–or a group with which they identify–will get more in return than they are asked to give in the first place.*

This is exactly the message conveyed by surveys that appear professional and address important problems that matter to respondents—in a nutshell, what this chapter is all about.

Mail Surveys

Overview

Imagine you have just come home from work and have a minute before putting groceries away and starting dinner. When checking your mailbox, you find two duplicate catalogs, assorted bills, a newspaper, and a mimeographed letter from your congressional representative. Also in the mail is an envelope with bulk-rate postage addressed to you with a label, last name first. Does it strike you as a letter that should be opened and taken seriously? Probably not. That is why even *envelopes* merit attention in successful mail surveys.

The decennial census is one survey in which questionnaires are distributed by mail. On March 23, 1990, the U.S. Census Bureau sent an individually addressed questionnaire to almost every household in the entire country. They hoped to get as many people to respond as possible and so conducted an intense, nationwide publicity campaign.

By April 14, only 63 percent of all households had completed and returned their questionnaires—a decent response rate for many mail surveys but not good enough for the decennial census (Kulka et al. 1991). To learn what could be done to improve response rates in the year 2000, Census Bureau researchers followed up with a sample of people who had not responded in 1990. It turned out that some didn't even remember getting a questionnaire. (In other words, they didn't realize it had come in the mail and needed their attention.) Others opened the envelope and then threw it out. Some started filling out the questionnaire, but then became confused and stopped. And finally, some answered all the questions on the form, but never mailed it back to the Census Bureau.

A lesson from the 1990 Census is that one needs to get people's attention all the way through the process from opening a mail questionnaire to the point when it is mailed back. Experience with hundreds of mail surveys over the years has convinced us that very high response rates are possible by using personalized correspondence, repeated mailings, and stamped return envelopes. We recommend a *basic* survey procedure that includes at least four separate mailings:

- *First:* To all members of the sample—a personalized, advance-notice letter. Its purpose is to tell people they have been selected for the survey and they will be receiving a questionnaire.
- *Second:* About one week later, again to all members of the sample—a personalized cover letter with slightly more detail on the survey, a questionnaire, and stamped return envelope.
- *Third:* Four to eight days after the questionnaire goes out, again to all members of the sample—a follow-up postcard thanking those who have responded and requesting a response from those who have not.
- *Fourth:* Three weeks after the first questionnaire goes out, to those who have not yet responded—a new personalized cover letter informing people, "We have not yet heard from you," with a replacement questionnaire and stamped return envelope.

For the general public, this four-mailing, basic procedure should yield a 50 to 60 percent response rate. More specialized populations are likely to respond in higher proportions. If study objectives make a higher response rate necessary, it is necessary to increase the number and type of contacts used in the basic procedure. We offer specific suggestions on how to do this later in the chapter.

Regardless of how many contacts are made, it is essential that they are timed for maximum effect. Waiting too long between mailings makes it likely that people will forget or even discard material they've already received. People often open their mail, put some pieces on a shelf, and throw them out after a week or so. That's why it is important to send the first three mailings close together—to interrupt the garbage cycle from shelf to wastebasket. If anything, time between the mailings should be shortened rather than lengthened.

The process of repeated, personalized, and well-timed contacts is designed to send one basic message, which is that each respondent's participation is essential to the success of an important study. Their personal knowledge about the issue being studied is critical to solving a problem that affects them or people they know directly, for example, a tax increase or the quality of children's education. Communicating this message is the key to getting high response rates and quality responses. It depends, at the minimum, on carrying out each step of the basic procedure outlined above.

Ahead of Time

Evaluate how much help you need to do the survey. One reason people choose the mail method over its alternatives is because mail survey tasks can fit around other work. For example, without disrupting his or her schedule too much, a school secretary can provide the clerical assistance necessary for a group of volunteers to survey parents at an elementary school. Similarly, the president of a small business can survey potential customers with just one clerical assistant who has been temporarily freed from other office responsibilities.

Remember, though, the timing of each mailing is critical. There will definitely be days when mailing lists and cover letters are top priority, preempting all other work. If normal demands on the people doing the survey cannot accommodate this disruption, arranging for additional volunteer or paid help may be necessary.

Transfer the sample list to a computer. The clerical work involved in a mail survey is greatly reduced when the sample list is loaded onto an office computer, because of the repeated tasks involved in the mailings. For example, respondent names and addresses are printed out at least five times (on cover letters, outgoing envelopes, and follow-up postcards). Also, the mailing list is sorted between the third and fourth mailings to cull respondents who have already responded and do not need another letter. These tasks can be done far more efficiently if the sample list is loaded on a computer than if typewriters are used for the whole process.

Publicize the survey. In some cases, announcing a survey publicly helps to legitimize it. If a growers' association surveys raspberry producers about their use of pesticides, an announcement about the survey in the association newsletter may help convince skeptical respondents that the survey will be useful. If a housing task force surveys residents of a small community on the sensitive subject of income, a supportive article in the local newspaper may alleviate concerns about confidentiality. These are examples of mail surveys in which good, advance publicity may affect response rates because respondents are likely to get positive information about the survey before being asked to respond. Of course, negative coverage can have the opposite effect, so advance publicity should be handled carefully.

While publicity sometimes legitimizes a survey, it is no substitute for follow-up contacts. As we explain below, the most effective follow-ups are always personalized.

Prepare as much ahead of time as possible. An advantage of mail surveys is that some of the critical work can be done ahead. For example, envelopes for the first mail-out can be addressed, letters drafted, envelopes stuffed and follow-up postcards prepared. This advance work leaves people working on the project free for other tasks that must be done during the survey itself, such as responding to inquiries and promptly checking names off the mailing list as questionnaires come in.

We strongly encourage people to prepare follow-up postcards at the same time the second mailing is organized. Too often, people wait to work on the postcard until after they send out the second mailing. Then, any problems with printing or supplies make it impossible to complete these two contacts within four to eight days of each other.

The Mailings

First mail-out. The first contact with respondents is made with a personalized but very short advance-notice letter sent to all members of the sample. The purpose is to get people interested enough to open the questionnaire when it comes. The letter should

- tell people the questionnaire is coming;
- briefly state why the survey is being done; and
- explain that participation will be greatly appreciated.

Figure 8.1 shows an example of a brief advance-notice letter that effectively communicates each of these three points.

As with all correspondence in a mail survey, the letter should have a date and handwritten signature. To get people's attention, we recommend sending it in a business-size envelope, with name and address individually typed or printed, and using first-class—*not* bulk-rate—postage. The advantages of first-class postage are threefold: It gets mail through the postal system faster, it's forwarded or returned

Figure 8.1

Advance-notice letter sent to all members of the sample selected for a mail survey and, with minor changes, for a drop-off survey

Washington State University

Social and Economic Sciences Research Center

Pullman, WA 99164-4014
(509) 335-1511
FAX (509) 335-0116

September 24, 1994

Mr. K. Seale
4522 7th Street
Colfax, Idaho 83855

Dear Mr. Seale:

Within the next few days, you will receive a request to complete a brief questionnaire. We are mailing it to you in an effort to learn how residents of Pima County feel about the use of alcohol in their community.

The survey is being conducted to better inform legislators, school authorities, and others who must make decisions related to alcohol consumption.

We would greatly appreciate your taking the few minutes necessary to complete and return your questionnaire.

Thank you in advance for your help.

Sincerely,

Priscilla Salant
Project Director

if necessary, and it looks important. Mail with bulk-rate postage is slower, gets discarded by the Postal Service if addressed incorrectly, and many people don't hesitate to throw it out if unopened.

Second mail-out. The second contact in a mail survey is made about one week after the advance-notice letter. It includes a personalized cover letter with a handwritten signature, questionnaire, and business-size, stamped return envelope with postage. As with the advance-notice letter, we recommend sending these materials in a business-size envelope, with names and addresses individually typed (or laser-printed), and a first-class stamp on the return envelope.

The cover letter that accompanies the questionnaire is critical to making the case that potential respondents should take the survey seriously. Figures 8.2 and 8.3 illustrate two very different cover letters. The first seems disinterested and almost perfunctory, suggesting that whether the respondent replies is of little consequence. The second is more personalized yet businesslike. Hence, it more successfully communicates the message that each respondent is important enough to warrant individual attention.

Note that the cover letter in Figure 8.3 mentions an **identification number**. We recommend stamping this number on the front

Figure 8.2
A mail survey cover letter that is unlikely to produce a response

Washington State University

Social and Economic Sciences Research Center

Pullman, WA 99164-4014
(509) 335-1511
FAX (509) 335-0116

Dear Pima County Resident,

Enclosed please find a questionnaire that is being sent to a random sample of all county residents. It concerns issues related to alcohol use in our community. We would appreciate your help in addressing these issues.

Thank you for your time and attention.

Sincerely,

Priscilla Salant

Washington State University

Social and Economic Sciences Research Center

Pullman, WA 99164-4014
(509) 335-1511
FAX (509) 335-0116

September 28, 1994

Mr. K. Seale
4522 7th Street
Colfax, Idaho 83855

Dear Mr. Seale:

As a resident of Pima County, you may have heard about problems related to the use of alcohol in your community. Knowing how people view the importance of these problems—and what kinds of solutions should be considered—is vital to local and state legislators, school authorities, and others who must make decisions about alcohol consumption issues.

Your household is one of the small number in which people are being asked to give their opinion on these matters. It was drawn randomly from a list of all registered voters in Pima County. In order that the results of the study truly represent the thinking of people in your community, it is important that each questionnaire be completed and returned in the envelope provided.

You may be assured of complete confidentiality. The questionnaire has an identification number for mailing purposes only. This is so that we may check your name off the mailing list when your questionnaire is returned. Your name will never be placed on the questionnaire itself.

I would be happy to answer any questions you may have about this study. Please write or call me collect at 509/335-1511.

Thank you very much for your assistance.

Sincerely,

Priscilla Salant
Project Director

Figure 8.3

A mail survey cover letter that successfully communicates the importance of individual responses

of each questionnaire and a corresponding number next to the appropriate name on the mailing list. This system lets us keep track of who has returned their questionnaire and who needs a follow-up mailing. It cuts costs because we avoid sending reminders and duplicate questionnaires to people who have already responded. Having an identification number on questionnaires appears to have minor effects, if any, on response rates.

Sometimes the person to whom the letter is addressed is not the one who should respond to the questionnaire. This is often true when household lists are used to draw samples. In that case, we add to the letter a statement like "In order for our results to represent all of the people in this state [or whatever population is of interest], we ask that the questionnaire be completed by the adult in your household who had the most recent birthday."

The last enclosure in the mail-out package is a preaddressed, return envelope, complete with postage. We strongly advise against expecting respondents to pay for postage when they return their questionnaires because it significantly increases what is demanded of respondents.

We prefer to put stamps on each return envelope that is mailed out. This practice has been shown to increase response rates by 2 to 4 percent over rates obtained with business reply envelopes. It also encourages respondents to return their questionnaires more quickly. Business reply envelopes require a permit from the post office. For organizations that already have a permit, these envelopes offer a small cost advantage insofar as postage is only paid when a respondent returns the questionnaire.

Third mail-out. Four to eight days after the questionnaire is mailed, a postcard follow-up is sent to all members of the sample. The purpose is to thank those who have responded and nudge those who have not. Overall, the basic message is the same—the respondent's

Figure 8.4
Postcard follow-up for all members of the sample, one week after the first questionnaire is mailed

Last week, a questionnaire seeking your opinions about alcohol use was mailed to you. Your name was drawn randomly from a list of all households in Pima Counry.

If you have already completed and returned the questionnaire to us, please accept our sincere thanks. If not, please do so today. We are especially grateful for your help because we believe that your response will be very useful to policy makers and public officials.

If you did not receive a questionnaire, or if it was misplaced, please call us collect at 509/335-1511 and we will get another one in the mail to you today.

Sincerely,

Priscilla Salant
Project Director
Social and Economic Sciences Research Center
Washington State University
Pullman, WA 99164-4014

personal knowledge about the issue being studied is critical to solving a problem that affects them directly. Again, the tone is businesslike and the style is personalized. Figure 8.4 illustrates how such an effect might be achieved.

Fourth mail-out. Three weeks after the second mail-out, a new personalized cover letter with a handwritten signature, questionnaire, and preaddressed return envelope with postage are sent to all members of the sample who have not yet responded. The purpose is

Figure 8.5
Cover letter included in the fourth mail-out, designed to encourage people who have not responded to do so

Washington State University

Social and Economic Sciences Research Center

Pullman, WA 99164-4014
(509) 335-1511
FAX (509) 335-0116

October 19, 1994

Mr. K. Seale
4522 7th Street
Colfax, Idaho 83855

Dear Mr. Seale:

About three weeks ago, we wrote to you seeking your opinions about issues related to alcohol use in Pima County. As of today, we have not received your completed questionnaire. We realize that you may not have had time to complete it. However, we would genuinely appreciate hearing from you.

The study is being conducted so that citizens like you can affect policies related to alcohol use. We are writing to you again because the study's usefulness depends on our receiving a questionnaire from each respondent. Your name was drawn through a scientific sampling process in which every registered voter in Pima County had an equal chance of being selected. In order for information from the study to be truly representative, it is essential that each person in the sample return their questionnaire.

In the event that your questionnaire has been misplaced, a replacement is enclosed. We would be happy to answer any questions you have about the study. Please write or call me collect at 509/355-1511.

Sincerely,

Priscilla Salant
Project Director

to attempt once again to convince people they are individually important to the study. The tone of the cover letter is more insistent but still businesslike, as illustrated in Figure 8.5. The single most important line in the letter is "As of today, we have not received your completed questionnaire." This is a strong, personalized message that tells people someone is waiting for *their* particular response. It is a powerful incentive to complete the questionnaire and return it as soon as possible.

The timing of the fourth mail-out is critical. We recommend against putting it off in the hope that more questionnaires will trickle in and, therefore, that the size of the fourth mailing will be smaller. It is far more advantageous to recontact people before they forget the previous mailings. Again, this conveys the message that the respondent and the study are important.

When a Higher Response Rate Is Necessary

All other things being equal, response rates increase when people are contacted more often. For some surveys, the 50 to 60 percent response rate that comes from making four contacts may be sufficient. In other cases, however, project objectives cannot be achieved without a higher response. Several options are available when that is the case.

The first option is to give potential respondents a monetary incentive to complete and return their questionnaire. Research has shown that when combined with the multiple contact strategy discussed above, an incentive as small as one to two dollars mailed with the first questionnaire can increase response rates 5 to 8 percent (James and Bollstein 1990). In a mail survey of construction company owners, larger incentives of $5 to $20 increased response rates even further. Rates did not vary according to whether incentives came in the form of a check or cash. A promised incentive (in other words, one that would be mailed after the questionnaire had been returned) had a minimal impact on response rates (James and Bollstein 1992).

The second option is to call nonrespondents a few days after mailing the replacement questionnaire. You might say something like "Other people have had questions about the survey and we wondered if you might also." The idea is to communicate personally with people and to inquire politely if they need help with or have concerns about the survey. You can then send a replacement questionnaire if necessary. We have found that such calls help to identify ineligible respondents from the sample list.

Finally, a fifth mailing can be sent by two-day priority mail, which costs $2.90 per questionnaire as this book goes to press. The large, priority mail (Postal Service) envelope is boldly printed in red, white, and blue. People who have not yet responded may be convinced of the project's importance as conveyed by the nature of the envelope and the extra expense that survey organizers incur with two-day service.

Each of these additional contacts is a means of increasing response rates in a way that is not likely to offend or put undue burden on the respondent. Each can be used separately or in conjunction with the others. The highest response rate will be achieved when a prepaid financial incentive is used *and* an additional contact is made by either telephone or two-day priority mail.

If You Have Addresses but No Names

A recently developed technique for general-public mail surveys may help overcome the problem of address lists that don't include names. Such lists are often available from local governments or utility companies.

Like the basic procedure discussed above, this technique involves an advance-notice letter followed a few days later by the questionnaire and, in another few days, by the reminder postcard. Since none of this communication can be addressed to a specific person, the one who has had the most recent birthday is asked to complete the questionnaire. The pre- and postquestionnaire mailings seem to encourage people to open the correspondence despite the absence of a name. The success of this technique appears to depend on the population surveyed (Krysan et al. 1993).

Final Comments on Mail Surveys

We have made two key points in this section. First, response rates in mail surveys depend very much on the number of contacts. There is no substitute; more contacts mean higher response rates. The goal is to make repeated and well-timed contacts in a pleasant, inoffensive way that will further encourage response. We recommend (at the minimum) a basic four-mailing method that is designed to convince each respondent that he or she is important to the study's success. Communicating this message significantly increases the chances of getting a high response rate and quality responses.

Second, mail surveys require advance planning. Make sure you have professional-looking materials, adequate help, and enough money to carry out each stage of the survey. Omitting any individual step of the basic procedure can have a major impact on the success of your survey.

In the final analysis, what really matters is the overall look and feel of a mail questionnaire. We find it helpful to draw an analogy with the look and feel of a new car. Undoubtedly, people shopping for a new automobile consider some of the component parts. Some look for high gas mileage, others for air-conditioning, and still others for air bags. Even so, it is the total effect that usually determines whether a car is purchased or not. The industry spends tremendous amounts of money getting the components to work together and designing an overall appearance that appeals to consumers.

Understandably, many people focus on individual components when they design a mail questionnaire, evaluating the importance of each question and design feature. Sometimes they fail to go the extra mile. They don't look at how the questionnaire appears as a whole. They forget to examine whether the parts fit together, for example, whether transitions are smooth and logical, whether the layout is uncluttered and directions easy to follow. Even the most well-planned implementation procedure can't compensate for oversights like these. Even financial incentives, first-class stamps, and multiple contacts can't redress an unattractive or illogical questionnaire.

For all these reasons, we urge people to pay attention to the whole survey process. Don't pick and choose features you personally like or find convenient. Instead, integrate all the elements together, from writing the first question to sending the last reminder.

Readers who are familiar with the "total design method" for mail surveys (Dillman 1978) may wonder why the procedures presented here are not exactly the same as in the earlier book. While retaining our overall perspective that people respond when they think the benefits outweigh the costs, we've presented less detail here and have also tried to update our recommendations. For example, based on recent research, we now recommend using financial incentives to improve response rates (Dillman 1991). To make multiple contacts more acceptable, we also suggest using an advance-notice letter rather than a second replacement questionnaire. Finally, we suggest slightly different or targeted methods for certain populations.

Telephone Surveys

Overview

If you have chosen to do a survey by telephone, you probably made your decision for one or more of the following reasons: You are in a hurry to get results, you want better control over which person in the household or business answers your questions (compared to a mail survey), your topic lends itself to discussion by phone rather than in writing, or face-to-face is too expensive. These are good reasons to choose the telephone method. Good phone surveys *can* be done quickly and inexpensively, it is possible to control who responds, and accurate information is possible to get over the telephone. However, because the telephone method is demanding and complex, the project supervisor must know what he or she is doing and understand how and why telephone surveys work.

For telephone surveys to be successful, everything must come together efficiently and quickly once interviewing begins. Questions always arise the minute dialing begins. What happens if someone in the sample is out of town for two weeks? Should they be called back or dropped from the sample? Do the interviewers call an operator if a number from the sample has been disconnected? What happens if a respondent wants to talk "to whomever is in charge"? Will anyone be around to interview a respondent who is only available at 6:30 A.M. on Sunday? What about answering machines? A knowledgeable supervisor must provide interviewers with immediate answers to questions like these in order to preserve the integrity of the sample and collect high-quality data.

Planning ahead is essential. A place to conduct the phone interviews must be arranged, survey materials prepared, and then interviewers hired and trained. There is another, larger issue that has to do with how hard it is to conduct telephone surveys in the 1990s. People are besieged by phone calls from charities and telemarketers. Most of us have had our dinner interrupted more than once by someone asking for a donation or trying to sell us something. And most of us, more than once, have hung up abruptly. A point we made at the beginning of the chapter is especially important in telephone surveys: You *must* convey professionalism and legitimacy to get good response rates.

Ahead of Time

Arrange for facilities and equipment. The most successful telephone surveys are conducted in one central location set up so that interviewers and their supervisor can work together. This arrangement is well suited for dealing with the unexpected problems that arise in every telephone survey.

The alternative may sound tempting. Why not ask interviewers to make phone calls from their home or office? Wouldn't this avoid the cost of setting up a central office? In our experience, the benefit of a central location far outweighs the cost for three reasons.

First, a centralized arrangement allows the supervisor to monitor and support interviewers, especially those with little experience. Second, it makes it possible to conduct phone surveys quickly. Interviewers never finish their share of questionnaires at the same time—some get a string of not-at-homes and ineligibles while others are busy completing interviews. As long as all the calls are made from one place, the supervisor can redistribute uncompleted questionnaires to

How Many Telephones and How Many Interviewers?

Our rule of thumb is that if a phone interview takes 20 minutes to complete, an experienced interviewer can average one every 40 minutes under ideal conditions. To be on the safe side, we budget 50 minutes per 20-minute interview. This allows time for "no-answers" and "respondents not at home," as well as for going back over the questionnaire to check for completeness and legibility.

Let's assume you want to end up with 200 completed interviews. That means you need about 167 hours of interviewing time (200 interviews × 50 minutes ÷ 60 minutes/hour). If you have six phones in use for four hours every evening, it will take you seven days to get 200 completed questionnaires.

Keep in mind that toward the end of every phone survey, it gets harder and harder to find people at home. (The respondents who are easiest to reach get interviewed first, leaving those who are hardest to reach for later.) If you want to keep your interviewers busy, you'll need fewer phones in use as you get closer to the end of the survey.

Telephone interviewing is a demanding job that takes concentration and consistent good humor. Few interviewers can work night after night for an entire survey. To give interviewers the breaks they need, we recommend hiring roughly one-and-a-half times as many people as there are telephones.

Realistically speaking then, if you want to complete 200 twenty-minute interviews over the course of seven days, you need 6 phones and 9 interviewers.

What about Computer-Assisted Telephone Interviewing?

Interest in speeding up and standardizing data collection on telephone surveys has led to the use of computer-assisted telephone interviewing, or CATI, for short. In a centralized CATI system, each interviewer sits at a sound-protected work station equipped with a computer monitor, telephone, and headset. Instead of doing interviews with pencil and paper, he or she works with the computer. It may generate a telephone number for the interviewer to dial, and it then prompts him or her through the entire interview process. Whenever necessary, the computer also branches through the appropriate sequence of questions and lets the interviewer know when he or she enters an out-of-range or inconsistent response.

CATI enables researchers to complete complex surveys in record time because data can be edited and analyzed almost instantly. For this reason, CATI systems are increasingly being used by marketing and polling firms, government agencies, and universities. These organizations conduct multiple surveys, so they can often afford the high cost of setting up a CATI system.

Most people who read this book will find that CATI is far beyond their means. However, if you have the option of contracting with an organization set up to conduct CATI, you can learn more about the procedure from Frey's excellent discussion of the subject in *Survey Research by Telephone* (1989) and *Telephone Survey Methods* (Lavrakas 1993).

those who finish ahead of schedule. This way, everyone keeps busy and uses their time efficiently. Third, and just as important, interviewers usually have more fun working together than all by themselves.

Although it is a good idea to have interviewers working in one central place, they don't necessarily need a facility that has been custom-designed for phone surveys. A temporary arrangement can be set up with leased phone lines installed in a room normally used for another purpose. (Some phone companies are willing to make such an arrangement for minimal or no cost.) Another alternative is to borrow telephones in employee offices at a college or agency. This is most practical for general-public surveys in which most calling is done in the evening and will not interfere with normal business hours.

If possible, interviewers should be able to work comfortably, without being disturbed by others, but within close range of their supervisor. It is also very useful to have a large work table in the middle of the room for sorting questionnaires according to whether they are complete or need further attention.

Prepare the survey materials. A complete set of materials for telephone surveys consists of

- an advance letter *if* names and addresses are available;
- the questionnaire, with a cover page for identification information and call record; and
- help sheets for the interviewer, including general information on interviewing techniques and specifics about the survey.

An excellent way to counter people's reluctance about telephone surveys is to send **letters in advance** of actual calls. Of course, this is only possible if you have names and addresses ahead of time, as when a sample is drawn from telephone directories. (Even then, you may have incomplete names and will need to omit the salutation from your letter.)

Although good advance letters are short, they serve several important purposes. (See Figure 8.6.) First, they decrease the element of surprise that comes from unexpected telephone calls. Second, they provide legitimacy by introducing the survey and distinguishing it as a genuine research effort. (This is especially important since so many sales pitches are disguised as telephone surveys.) Third, they explain the within-household selection process, if one is used. Fourth, they let respondents know when to expect a call and that a more convenient time can be arranged if necessary. Finally, advance letters thank respondents ahead of time for participating in the survey.

When mailing advance letters is not possible, another way of letting people know they might be called is to announce the survey publicly. The purpose of advance publicity is to let people know the project is legitimate and to reduce the element of surprise.

The second—and central—part of survey materials is the **questionnaire,** which was described in detail in Chapter 7. The only addition that needs to be made is a cover page to identify the respondent (by number and/or name) and to record information about each

Figure 8.6

An advance letter used in a telephone survey

Washington State University

Social and Economic Sciences Research Center

Pullman, WA 99164-4014
(509) 335-1511
FAX (509) 335-0116

September 28, 1994

Ms. F. Goldmark
3145 Main Street
Palouse, Washington 99134

Dear Ms. Goldmark:

Within a week or so, we will be calling you from Pullman as part of a research study. This is a state-wide survey in which we are seeking to understand how Washington residents feel about the communities in which they live and what should be done to improve them.

We are writing in advance of our telephone call because we have found that many people appreciate being advised that a research study is in progress and that they will be called.

When our interviewer calls, she (or he) will ask to interview an adult member of your household. In order that our results represent all of the people in Washington, we ask to interview the adult who has had the most recent birthday.

Altogether, the interview should take about ten minutes. If by chance we call at an inconvenient time, please tell the interviewer and they will be happy to call back later.

Your help and that of the others being asked to participate in this effort is essential to the study's success. We greatly appreciate it.

If you have any questions, please don't hesitate to ask our interviewer. You may also call me collect at (509) 335-1511 or contact me by mail.

Cordially,

Priscilla Salant
Project Director

telephone contact. (See Figure 8.7.) It's very likely that *at least* several calls will be necessary to complete many of the interviews and furthermore, that different people will be making the calls. Using a call record is the best way to keep track of what has happened with previous calls and when someone should try completing the interview.

The third document that should be prepared ahead is a set of **help sheets**. Help sheets are used during interviewer training and also once the survey begins. They include specifics about the survey as well

Community Futures Survey

WSU Department of Agricultural Economics

Q1 Household/respondent ID# _______

Q2 Area code and number ( ) ___ - _______

Call Record					
Call no. and interviewer	Date	Time	Contact		Result Code and Comments
			Resp.	Other	
1.					
2.					
3.					
4.					
5.					
6.					
7.					
8.					
9.					
10.					

Additional Comments:

RESULT CODES

Respondent not contacted:

01 Answering machine
02 Busy
03 Ring; no answer
04 Number not in service
05 High-pitch screech (i.e., fax)
06 Not in sample (e.g., nonresidential)
07 No adult at home
08 Respondent identified; callback needed; appointment made
09 Respondent identified; time not specified for callback

Respondent contacted but interview not completed:

10 Appointment made to call back
11 Callback should be made; time not specified
12 Refusal

13 *Respondent contacted; interview completed*

14 Other

Figure 8.7
Sample cover page for a telephone questionnaire

as general information on interviewing techniques. The first sheet is one that can be posted at each workstation. (See Figure 8.8.) It is intended to help interviewers answer questions that respondents commonly ask. Other help sheets briefly summarize the basic principles of proper interviewing techniques and explain how to use the telephone, keep a call record, file material at the end of the day, and maintain confidentiality. Sample help sheets for a telephone survey are included in Appendix 8.A.

Figure 8.8

Information for telephone interviewers to help answer respondent questions

What the Respondent May Want to Know about the Survey

I. **Who is sponsoring the survey?**
This survey is being sponsored by the Department of Agricultural Economics at Washington State University.

II. **What is the purpose of the study?**
The survey is being conducted to assess quality of life issues and local concerns of residents in this community. Information will be collected on such things as city services, housing, health care needs, and economic development.

III. **Who is the person responsible for the survey?**
Priscilla Salant from the Department of Agricultural Economics is responsible for managing this study. Her phone number is 335-7613.

IV. **How many people will be participating in the survey?**
We will be attempting to complete 1000 interviews.

V. **Who are you? Who is conducting the survey?**
I am a student (or resident of Pullman, WA) working part-time for the Department of Agricultural Economics at Washington State University.

VI. **How did you get my name?**
The names for this study were selected randomly from lists of households located in this community.

VII. **How can I be sure that this is authentic?**
I would be glad to give you our telephone number here at WSU in Pullman, WA, and you may call my supervisor. The evening supervisor is Susan Lamont and the daytime supervisor is Bob Anderson. They can be reached by telephoning 335-1511.

VIII. **Is this confidential?**
Yes, most definitely. After the research is completed, the answers are put on a computer without names, addresses, or any means of identification. All of the information that is released is presented in such a way that no individual response can ever be traced.

Also, the matter of confidentiality is terribly important to the success of our research center because we conduct many surveys. Therefore, we are very careful to protect people's anonymity.

IX. **Can I get a copy of the results?**
You may contact Priscilla Salant, at the Department of Agricultural Economics. Her telephone number is (509) 335-7613.

X. **What will the results be used for and how will the study help me?**
The information will be used to set city policies and identify areas where improvements need to be made. By learning more about the opinions of residents like you, we hope the project will contribute to making our community a better place to live.

Choose and train the interviewers. It is unlikely that many of you will have a ready supply of experienced telephone interviewers waiting to work on a survey. However, finding people who can be trained and are at least temporarily available should not be difficult. Neither the skills nor the hours required are extremely limiting. Figure 8.9 shows how a telephone interviewer's job description might look.

Whether they volunteer or are paid, potential interviewers should be judged on their ability to read questions fluently and communicate verbally. Good telephone interviewers can read questions out loud without hesitating or stumbling over words. In addition, they have clear and pleasant telephone voices that do not interfere with those of other interviewers working nearby. Finally, they can respond quickly to respondents' questions without losing composure or interrupting the flow of the interview. The best way to learn whether people have these qualities is to let them study part of a questionnaire and then listen on another extension as they conduct a trial interview over the phone.

Training the interviewers is not optional! Good training sessions are always interactive. People need ample opportunities to practice their interviews out loud. This gives everyone a chance to learn from others' mistakes and improve their own ability to conduct interviews. Remember that telephone surveys are verbal, above all else. You cannot expect fluency without practice.

In addition to learning about the questionnaire, new interviewers need instructions on many other aspects of the project. For example, they need to know

- background information about the survey;
- the basics of proper interviewing (see Appendix 8.B);
- how to operate the telephone and complete a call record; and
- how to edit and file completed questionnaires.

If interviewers are paid employees, they also need to know administrative details regarding time sheets, IRS forms, and the like.

Schedule the interviews. When interviewers should be scheduled to make their calls depends entirely on whom is being surveyed. The key is to figure out when people are most likely to be available and willing to be interviewed. For work-related surveys, you should probably call during business hours. (There are obvious exceptions, as in the case of people who don't have a telephone at work or would be unable to speak freely unless in their own homes.) For general-

How Many Calls Should Be Made to Each Respondent?

The more calls you make, the greater the chance you'll be able to contact the respondent. For some organizations that do national surveys, making 20 callbacks is standard procedure. However, in local surveys for which budgets may be smaller, try making six calls to each person in the sample. If your response rate is lower than your goal, call one or two more times. If someone is consistently out at a particular time of day, try calling earlier or later.

Figure 8.9

A telephone interviewer's job description

Telephone Interviewer

A telephone interviewer is responsible for conducting telephone interviews for survey research purposes. Telephone interviews are primarily conducted during evening hours (5:00 to 9:00 P.M.), Sunday through Thursday.

Before conducting telephone interviews with actual respondents, an interviewer must:

a) Attend the "Basics of Proper Telephone Interviewing" training session.
b) Complete project-specific training prior to conducting any interviews.
c) Complete a series of practice telephone interviews and be subjected to monitoring and evaluation by a trained staff member.
d) Sign a statement of confidentiality pertaining to telephone interviews.

The duties involved in conducting telephone interviews include:

a) Contacting potential respondents over the telephone. This will include calling without prior contact and scheduled appointments.
b) Persuading respondents to participate in the survey.
c) Using project-specific information to answer a respondent's questions and objections.
d) Applying all rules and standards contained in the Interviewing Training Manual in a consistent and reliable manner.

All interviewers must:

a) Be able to communicate effectively over the telephone.
b) Be courteous and pleasant with respondents.
c) Have the ability to respond to difficult situations in a mature and professional manner, while maintaining the integrity of the data.
d) Demonstrate ability to follow directions.
e) Be conscientious in work habits.
f) Demonstrate dependability in their work schedule.
g) Contribute to a positive work environment and interact well with supervisors and co-workers.

public surveys, we have found that weekday evenings and weekends are best for finding people home and willing to talk. We usually start the project by calling from 5:00 to 9:00 P.M., Sunday through Friday, with additional sessions during the afternoons and evenings on Saturdays and Sundays.

It is difficult at the beginning of a survey to schedule interviewers all the way through the project since the pace and success rate vary day by day. Also, as the project reaches the final third of the sample, fewer interviewers are needed each session. (See the shaded text, How Many Telephones and How Many Interviewers?) Hence, we suggest beginning a survey with an interviewer at each phone, but remaining flexible about how many will be needed as time goes on.

When the Interviews Begin

The big payoff from advance planning comes when interviewers begin placing their first calls. With a little luck, all the details come together at that point, so the survey proceeds smoothly and efficiently.

Unfortunately, even the best prepared team usually runs into unexpected problems. That's when an alert supervisor becomes essential. His or her job is to continuously monitor the interviewers' progress. Specifically, he or she should listen to interviews, talk to interviewers between calls, check questionnaires for completeness, and make sure incomplete questionnaires are handled correctly. (For example, callbacks need to be scheduled for respondents who aren't home when an interviewer calls, and disconnected numbers in the sample may need to be replaced with working numbers.)

Another of the supervisor's jobs is to decide how to handle refusals. The question is: Should some respondents who refuse to be interviewed be recontacted? The answer is: It depends. If someone makes a conscious decision not to participate after hearing what the survey is about, then we recommend against trying to change their mind. On the other hand, if someone hangs up right away without hearing the reason for the survey, we may decide they should be called back a few days later. The call could begin something like this:

> One of our interviewers attempted to call your household the other night. We may have called at a bad time so I wanted to check back with you. The survey we are doing is a very important one—we're trying to get people's views on handling the economic problems we face in our community. We would really appreciate hearing your opinion on these issues.

This process is called **refusal conversion.** It is almost always done on big national surveys but is less common on local projects. Someone doing a survey is understandably more sensitive to respondents in their own community who may be offended by persistent (though polite) interviewers. Ultimately, this is a judgment call that only a supervisor and/or survey sponsor can make.

What about Answering Machines?

Sometimes surveys address highly visible issues and a message left on an answering machine will convince people to call back collect or on a toll-free number. In other surveys, people aren't likely to be motivated about responding. For example, in an election preference poll, leaving messages on answering machines probably would not increase response rates by very much.

Final Comments on Telephone Surveys

The key point to remember is that a telephone survey is a relatively complex undertaking. It requires you to be organized (in terms of equipment, interviewers, and materials, for example) and knowledgeable about a host of details. Once interviewers begin dialing, many particulars must come together. Otherwise the process can get bogged down, and various errors will begin to compromise data quality.

Face-to-Face Surveys

Overview

Higher labor costs and travel expenses mean that face-to-face surveys usually cost more than either mail or telephone surveys. If you have decided that the benefits of a face-to-face survey outweigh the expenses, you probably made the decision for one or more of the following reasons: You are doing a small local survey where costs can be kept low; you cannot get a good population list; or the people you want to interview will not or cannot respond accurately to another type of survey.

In terms of actually putting a survey in motion, the face-to-face and telephone methods are similar. Both require planning ahead with respect to sampling details, survey materials, and staffing. In one important way, however, face-to-face surveys are different: Interviewers are on their own once they are sent out to find residences and start knocking on doors. That means extra work in the planning stage to make sure the right people are recruited, trained well, and equipped with proper materials.

Ahead of Time

Prepare survey materials. More materials are required for face-to-face surveys than for other methods. A complete set includes

- an advance letter if names and addresses are available;
- an interviewer name tag;
- a letter from the survey director that explains the purpose of the project and that can be left with the respondent;
- an interviewers' manual, including general information as well as specifics about the survey;
- instructions to the interviewer about who is to be included in the sample and how they are to be located;
- the questionnaire with a cover page for tracking purposes; and
- visual aids for the interview, if appropriate.

Whenever possible, **advance letters** are used in face-to-face surveys for the same reasons as they are in telephone surveys. Most important, they decrease the element of surprise that comes from unexpected visitors. In addition, they provide legitimacy by introducing the survey and distinguishing it as a genuine research effort. They also explain the within-household selection process (if one is used) and, finally, thank respondents ahead of time for participating in the survey. We provide an example in Figure 8.10.

When addresses for people in the sample are unavailable, advance letters cannot be used. Even when letters are sent ahead, some will be mislaid and not read at all. Therefore, it is important to provide interviewers with some means of identifying themselves and the survey. An official identification card with color picture is desirable. We recommend providing individually printed project **name tags,** as well as a **signed original letter from the survey director** that explains the purpose of the project and that can be left with the respondent. If a newspaper article or other announcement about the survey has been published, interviewers can also carry a copy to show respondents.

Another important document that needs to be prepared for in a face-to-face survey is the **interviewers' manual**, preferably in the form of a three-ring notebook. The main reason to assemble an interviewers' manual is to give everyone involved in the survey a complete, detailed reference document on the project. In particular, interviewers can use the manual to prepare for their job and also to answer questions that arise while they are in the field.

Figure 8.11 shows the contents of a manual developed for a recent survey of farm families. Note that the manual included a wide range of background information on why the survey was being done, general interviewing techniques (see Appendix 8.B), and the survey site itself. It explains the survey form question by question, and it instructs interviewers on who is to be included in the sample and how they are to be located.

In our experience, the value of an interviewers' manual goes beyond its main purpose of helping staff people. Writing the manual gives us a reason to pull together in one place a document that tells the story of our survey. Manuals also help organize interviewer-training sessions and provide a reference on the project long after interviewing is over.

Another document that needs to be prepared is the **questionnaire**, which we discussed in Chapter 7. Sometimes interviewers carry a blank questionnaire that respondents may refer to during the interview. In any case, the questionnaire should include a cover page similar to the one used for telephone surveys. (See Figure 8.12.)

How Many Interviewers Are Needed for a Face-to-Face Survey?

The answer depends mainly on how much territory the interviewers are asked to cover. If each one is assigned 20 one-hour interviews in a two-block neighborhood, they might be able to complete three or four in a day. If each is assigned 20 one-hour interviews over 100 square miles, they'll be lucky to finish two a day at the beginning and less as the survey progresses.

Say you want to end up with 200 completed questionnaires. If interviewers complete three questionnaires each weekday, it will take 67 days, or about 14 weeks, of interviewing time. Under the best conditions, that means seven interviewers working for two weeks.

Toward the end of the survey, it gets harder and harder to find respondents at home. Therefore, a more realistic estimate is that it would take seven interviewers three weeks to get completed interviews.

Despite the extra labor costs, we strongly recommend that interviewers work in two-person teams whenever personal security is a concern. They can either conduct their interviews together or work in one small area at the same time.

Figure 8.10
Example of an advance letter for a face-to-face survey

Washington State University

Social and Economic Sciences Research Center

Pullman, WA 99164-4014
(509) 335-1511
FAX (509) 335-0116

October 7, 1994

Ms. R. Harrison
Route 3 Box 1334
Palouse, Washington 99134

Dear Ms. Harrison:

Within a week or so, we will be visiting you to arrange an interview as part of a research study. This is a survey of Washington farm families in which we are seeking to understand how the state's farmers are adapting to recent changes in Federal farm programs. The information will be used by USDA administrators to evaluate how program reform affects the well-being of our nation's farm families.

We are writing in advance of our visit because we have found that many people appreciate being advised that a research study is in progress and that they will be contacted.

When our interviewer comes to your farm, she (or he) will ask to interview the person who makes most of the administrative and managerial decisions for your operation. Altogether, the interview should take about forty-five minutes. If we arrive at an inconvenient time, please tell the interviewer and they will be happy to come back later.

Your help and that of other farmers being asked to participate in this project is essential to making the study a success. We greatly appreciate it.

If you have any questions, please don't hesitate to ask our interviewer. You may also call me collect at (509) 335-1511 or contact me by mail.

Cordially,

Priscilla Salant
Project Director

It includes space for interviewers to identify the household and respondent as well as to record information about the outcome of each contact. For complicated surveys in which various editors handle the questionnaire, we also recommend including information that allows a supervisor to keep track of editing and data entry.

In some surveys, interviewers use **visual aids** in conjunction with the questionnaire. For example, when asking about income levels, an interviewer might use a flash card such as the one illustrated in

Figure 8.11
Contents of an interviewers' manual for a face-to-face survey

INTERVIEWERS' MANUAL

Table of Contents

BACKGROUND **Page**

Figure 8.12

Questionnaire cover page for a face-to-face survey

1994 Washington Family Farm Survey

WSU Department of Agricultural Economics

Q1 Household/respondent ID# ____________

Q2 Address ____________

____

____

Call Record					
Call no. and interviewer	Date	Time	Contact Resp.	Contact Other	Result Code and Comments
1.					
2.					
3.					
4.					
5.					

Additional Comments (Use back of this sheet)

RESULT CODES

Respondent not contacted:

01 No answer
02 Not in sample (e.g., respondent moved)
03 No adult at home
04 Respondent identified; callback needed; appointment made
05 Respondent identified; callback needed; no appointment made

Respondent contacted but interview not completed:

06 Appointment made to call back
07 Callback needed; time not specified
08 Refusal

09 *Respondent contacted; interview completed*

10 Other

Office Use

	Name	Date
Field edit		
Office codes		
Office edit		
Data entry		

Figure 7.15, in the previous chapter. Visual aids often make sensitive questions less threatening, make interviewers more comfortable, and help respondents answer difficult questions.

Tell public officials about the survey. Let the police or sheriff's department know that you are doing a survey; let them know when interviewers will be making their rounds. You may want interviewers to carry an official letter that gives a phone number for people to call if they have questions about the survey's legitimacy.

Put together and train the staff. Face-to-face interviewers must have a variety of skills that are not required of people who conduct telephone surveys. First, they need logistical skills. For example, face-to-face interviewers must read maps, set up appointments, and keep expense records.

Second, face-to-face interviewers need good interpersonal skills. They should come across as friendly, direct, and honest, interested in what respondents say but not prying. Even their attire is important. Neat, modest clothing usually makes the best impression.

Third, face-to-face interviewers must be able to work independently—they are on their own once they start knocking on doors. After the survey is underway, they may have infrequent contact with the supervisor. And when the unexpected happens, they have to rely on their own resources.

In short, it takes a special and well-trained person to conduct face-to-face interviews. Therefore, hiring is more difficult and must be done more carefully than in telephone surveys, and training is especially critical. For this reason, most face-to-face surveys require at least a two-day training session that all interviewers are required and paid to attend.

The principles of conducting interviewer-training sessions for face-to-face surveys are similar to those for telephone surveys. First, it is important to schedule enough time to cover the entire questionnaire, *as well as* background on the survey, administrative details, and what to do with the questionnaire after an interview is completed. Second, the training session should be as interactive as possible. Instead of lecturing, training instructors should lead a discussion, encouraging questions from the students. Each interviewer should have several opportunities to participate in the interview, both as a respondent and an interviewer. The program should also involve one or more practice interviews with actual respondents.

Once the Survey Begins

Like the supervisor in charge of telephone surveys, the person who manages face-to-face interviewing plays a critical role. He or she should be readily available to field questions and solve problems that arise during the survey. In addition, he or she should have regular meetings with each interviewer, especially in the beginning of the survey. The reason for these meetings is to "field edit" the questionnaires, that is, to check for completeness, legibility, and overall consistency.

Field editing can make an enormous difference in data quality because it gives the supervisor a chance to catch recurrent errors on the interviewers' part. These errors commonly occur when an interviewer has misunderstood the purpose of a particular question or is reluctant to probe for sensitive information. Without conscientious field editing, it is entirely possible that a large proportion of questionnaires will be unusable for one reason or another.

Occasionally, an interviewer cheats by filling out a questionnaire without conducting the interview at all. Because interview assignments are typically clustered to gain efficiency, survey results are especially sensitive to such fabrications. Therefore, we strongly recommend asking for a telephone number during each interview so that 10 or 20 percent of all respondents can be recontacted to confirm that they have been interviewed. (It's a good idea to let people know that a supervisor may be calling sometime in the future.) By asking how long the interview lasted and double-checking a few of the questions, it's also possible to learn whether sensitive sections of the questionnaire might have been skipped.

Final Comments on Face-to-Face Surveys

Face-to-face surveys are the most complex to implement. They require the best trained staff, the most preparation, and the longest time. The most critical feature of face-to-face surveys is that interviewers work on their own—they may not talk to their supervisor for hours or even days at a time. For this reason, we recommend that readers who use the face-to-face method review their plans with a professional and experienced researcher before putting their survey in motion.

Drop-off Surveys

Overview

In a drop-off survey, questionnaires are personally delivered to members of the sample and then either collected or mailed back. This is a flexible method that can be adapted to different circumstances. For example, if volunteers are willing to put considerable time on the survey and make up for a low budget, they can deliver *and* collect the questionnaires. By personally communicating the survey's importance to intended respondents, these volunteers can help increase the response rate. (Similarly, if citizen involvement is a priority, the drop-off survey offers an excellent way to give people a chance to participate.) On the other hand, if volunteers are scarce and can't make up for a low budget, questionnaires can be left with someone other than the respondent or even at a house where no one is home.

One problem with the drop-off technique is that people sometimes don't pay attention to getting the questionnaires back. One common method is to go back to the household to get the questionnaires on a specified evening. This is a good idea, but it inevitably results in people either not being home or not having completed the survey form. Second and third callbacks sometimes become awkward when the questionnaire "still" isn't done.

To avoid these problems, consider two alternatives, both of which involve saving time and resources for the most difficult-to-obtain questionnaires. The first is to leave a stamped return envelope (but *not* a business reply envelope) and ask that the questionnaire be returned. Many respondents will send their questionnaires back with no reminder, and the rest can be recontacted.

The second alternative is to make the first callbacks in person on a scheduled evening. If people are not at home or haven't returned the questionnaire, leave a note and stamped return envelope. Some, but not all, questionnaires will be returned. People who do not respond can be recontacted.

Getting telephone numbers on the first visit, "in case we need to get in touch with you about the survey," may facilitate additional follow-ups.

The key point is that the drop-off method is flexible. It works by combining the best features of mail and face-to-face surveys—low cost, personal contact, and opportunities for follow-up. In the section below, we present several options for using the method, depending on various situations and information needs.

Are Names and Addresses Available in Advance?

The absence of names and addresses is much less of a problem with the drop-off method than with a mail survey. People working on the survey can actually go to the study area and systematically select a sample (every fourth household, for example). In other words, they can use a geographic area frame instead of a list. (See the discussion of area frames in Chapter 5.) At the same time, they can sample from within the household or business *and* get names and addresses for the purpose of sending reminder postcards.

If names and addresses *are* available ahead of time, advance-notice letters can be sent before questionnaires are dropped off. These letters alert people to expect a visit soon. The content is similar to advance-notice letters used in mail surveys. (See Figure 8.1.)

Whether or not letters are sent out ahead, it is a good idea to deliver questionnaires when people are most likely to be in. In a household survey, we recommend trying around supper time or during the day on weekends.

How Much Help Is Available to Do the Survey?

Having more help on a drop-off survey usually means getting a higher response rate. Handing each questionnaire to its intended respondent and personally explaining why the survey is being done is definitely the best delivery strategy. Of course, it is faster to leave the questionnaire with a responsible adult who can then pass it on to the respondent. If you have very limited help or cannot catch anyone at home, leave the questionnaire at the door, preferably in a transparent plastic cover. In any case, don't use mailboxes for drop-off or retrieval; they are legally reserved for the U.S. Postal Service!

Regardless of how questionnaires are delivered, we recommend including a cover letter. Even if you personally explain to each respondent that they should take the survey seriously, you need a good cover letter to which they can refer after you are gone. An example is illustrated in Figure 8.3.

What about Incentives?

In a neighborhood or small community survey, where people delivering questionnaires may know the respondents, a financial incentive might seem inappropriate. However, there are more acceptable (and creative) kinds of incentives. For example, you could encourage people to respond by giving them raffle tickets for a locally produced quilt or perhaps a free dinner for two at a nearby restaurant.

Final Comments on Drop-off Surveys

Drop-off surveys are a hybrid that can be adapted to a variety of information needs and local circumstances. The procedure that works best for you will depend on how much help you have and whether you use a list or area frame. If you choose this method, design your project so that it takes the best from mail and face-to-face surveys: good follow-up and personal contact. When used effectively, this combination results in a low-cost survey with good response.

Ethics, Regardless of Method

In this chapter, we have provided many specific suggestions on how to improve survey response rates. Readers may have gotten the impression that "anything goes" in survey research, as long as it helps convince people to respond.

On the contrary, we believe that as researchers we have clear ethical responsibilities. First, we must respect anyone who decides not to take part in a survey. We can try to convince them of the project's importance and of the value of their contribution; we can remind them if they forget about the survey; and we can offer incentives. But in the final analysis, we must stop before becoming coercive or offensive. This is probably hardest in surveys that involve contact by phone or in person because they are inevitably the most intrusive.

The second important ethical issue is that of confidentiality. Rare is the survey that does not include a statement to the effect that all responses will be kept confidential. Regardless of what is promised, however, the opportunity to betray information provided confidentially is almost always present. The reason is that most questionnaires carry identification numbers so that surveyors can keep track of who has responded.

We recommend that people take several steps to ensure confidentiality. First, make sure all project staff members understand their obligation to protect respondents' privacy. In a telephone or face-to-face survey, that means interviewers should never discuss any particular respondent with someone not related to the project. Regardless of the method, it also means that results will never be presented in a way that would reveal data from an individual respondent.

Second, make sure respondents understand if a survey is confidential rather than anonymous. Anonymous means that individual people *cannot* be associated with specific questionnaires; confidential means they *will not*.

Third, if questionnaire cover pages contain identifying information (such as name and address), remove and destroy the pages as soon as it is practical. If project staff members have no need to double-check data from a questionnaire or make another contact, there is no need to keep the cover pages.

Finally, keep only one master list that links respondent names with questionnaire identification numbers, and destroy the list when it is no longer needed.

This discussion has covered only the basics of ethical surveying. More detail is available from academic textbooks on survey research and from professional associations whose members conduct surveys.

Summary

Having read about how surveys are set in motion, you might sit back now and think about ways in which the personality of a project changes when implementation begins. Most obvious perhaps is that the tempo speeds up considerably. Before the point when data are actually collected, things usually proceed at a civil and relatively relaxed pace. For example, one can take time to develop the questionnaire. Adding new questions, revising others, and simply letting things cook awhile can improve the final product. In contrast, implementation has a much faster cadence. Once the process starts, deadlines must be met, mailings or interview sessions scheduled, and people organized. Failing to keep the pace can turn data collection into a shambles.

Another apparent change has do to with management. In many surveys, we've noticed that leadership shifts from one person to another as implementation begins. That is because the skills required in preparing for the survey are now less useful. Instead, strict schedules must be developed, lists made for who does what and when, and responsibilities assigned and monitored. Above all, attention to detail and the ability to get things happening on time are absolutely essential.

Finally, implementation usually signals the need for more people to work on the survey. It is possible for one person to produce a good questionnaire—from writing questions to producing a printed booklet. In contrast, survey implementation can seldom be completed without involving many people, each one responsible for one piece of the job.

This is the last of four chapters in which we have emphasized the core issues of survey design: sampling, question writing, questionnaire design, and survey implementation. Now is a good time to ask yourself if your survey is one that can be done by the methods we have outlined, or whether it demands more advanced procedures. For example, do you intend to conduct face-to-face interviews over a wide geographic area and therefore need to cluster interviews in a multistage design? Do you want to do a national telephone survey that involves random digit sampling and difficult-to-complete interviews? Or perhaps, a mixed-mode survey that requires starting with one method and then switching to another? If your intentions are this complicated, we strongly suggest that you use the list of references that appears below and get professional help.

On the other hand, if your survey seems to be within the scope of the methods we have outlined here, it is time to think about what to do with your data when the survey is over, the topic of Chapter 9.

For more detail on survey implementation see:

Standardized Survey Interviewing: Minimizing Interviewer-Related Error, by Floyd J. Fowler, Jr., and Thomas W. Mangione. Sage Publications, Newbury Park, CA, 1990.

Chapter 5, "Supervision I: Structuring Interviewers' Work" and Chapter 6, "Supervision II: Structuring Supervisory Work," in *Telephone Survey*

Methods: Sampling, Selection, and Supervision, by Paul J. Lavrakas. Sage Publications, Newbury Park, CA, 1987.

Chapter 5, "Implementing Mail Surveys," and Chapter 7, "Implementing Telephone Surveys," in *Mail and Telephone Surveys: The Total Design Method,* by Don A. Dillman. Wiley-Interscience, New York, 1978.

General Interviewing Techniques: A Self-Instructional Workbook for Telephone and Personal Interview Training, by Pamela J. Guenzel, Tracy R. Berckmans, Charles F. Cannell, Survey Research Center, Institute for Social Research, University of Michigan, Ann Arbor, 1983.

Appendix 8.A

Telephone Interviewer Instructions*

1. **Before you start.** Have these instructions and "What Respondents May Want to Know About the Survey" in front of you. You will also need a pen and some blank paper.

   Read over the instructions and the answers to respondents' questions to familiarize yourself with the procedures and with what to say in response to questions.

2. **Using the telephone.** To dial the telephone: For on-campus numbers (i.e., 335 prefix), dial 5-XXXX

   For Pullman numbers (i.e., 332 or 334 prefix) dial 8-XXX-XXXX

   For other long-distance calls, dial 6-(AREA CODE)-XXX-XXXX

   For information calls, dial 6-222-XXX-555-1212

   One beep per second means that the line is busy. Try again in a short while. Two beeps per second means that for some reason, probably mechanical, the call did not go through. Try again.

3. **Keeping a log of all calls.** The cover page stapled on the front of each questionnaire gives you a phone number and sometimes a name for each call you are to make. It also includes space for you to record information about your contacts. Fill in your initials and the date and time of the call. Also indicate to whom you talked (if anyone) and the correct result code.

4. **If you need help, excuse yourself and get a supervisor.** Sometimes respondents want to know more about a question or reasons for the study, etc., than you can tell them. If more information is warranted, in your judgment, don't hesitate to ask the supervisor for help.

5. **If the respondent becomes incensed, or uses abusive language.** Be nice! Do not hang up! Keep cool! (Such negative responses are not likely to happen.) If they do, be patient—maybe the person had a bad day. Some responses that might help:

---

* Adapted from interviewer-training materials used at the Social and Economic Sciences Research Center, Washington State University.

Yes, I see. Uh-huh.

Yes, I understand that you feel quite strongly about this matter. But we really do need the information.

If all else fails, call for a supervisor or wait for the opportunity to say something to this effect:

I think I can understand your feelings, and your not wanting to complete the interview. But, thank you very much anyway. Goodbye.

6. **When you hang up.**
   a. Record the time and enter the termination code for this interview.
   b. Make any corrections you need to for this interview.
7. **When you are done for the day.** Check out with the supervisor, explaining any callbacks that need special attention.

   Do not take anything home with you. All questionnaires, codebooks, etc., must remain in the office.
8. **After you have left.** We have an obligation to respondents to keep their interviews confidential. We feel very strongly that this obligation should be honored. Therefore, please do not tell anyone the substance of any interview or part of an interview, no matter how fascinating or interesting it was. Also, please avoid giving your own summary of findings. Just because 90 percent of your respondents feel a certain way does not mean that 90 percent of everyone else feels the same way. Confidentiality is essential. If you want a copy of results from this survey, let the supervisor know, and we will be sure you get them just as soon as they are available.

Appendix 8.B

The Basics of Proper Interviewing*

Main Points

1. Read each question exactly as it is written and in the order in which it appears in the questionnaire.
2. Read slowly.
3. Use standard feedback phrases for acceptable responses. Examples are "Thank you. That's important information" and "I see."
4. Use standard cues or probes to help the respondent give more complete answers to questions. Examples are "Could you tell me more about that?" and "Which would be closer to the way you feel?"

Two Types of Questions

1. **Close-ended or precoded questions** with predetermined response categories from which the respondent must choose.
2. **Open-ended questions** that ask the respondent to express reasons and feelings in his or her own words.

*Adapted from interviewer training materials used at the Social and Economic Sciences Research Center, Washington State University.

Question Wording

1. Surveys work only if everyone gets the same stimulus. Therefore, read questions exactly as they are worded in the questionnaire. Read questions with no additions, deletions, or substitutions.
2. Often you will find questions that contain parentheses. An example is "How well do you think your (husband/wife) understands you?" Make a choice from parentheses based on what you have learned about the respondent from the household listing and other questions.
3. Read the entire question before accepting the respondent's answer.

Question Order

1. Ask each respondent every appropriate question.
2. Don't skip a question because the answer was given earlier or because you "know" the answer.
3. In those situations in which the respondent has already provided information that probably answers the next question, you may preface the question with some combination of the following phrases:

   "I know we've talked about this," or "I know you just mentioned this, but I need to ask each question as it appears in the questionnaire."

   "You have already touched on this, but let me ask you . . . "

   "You've told me something about this, and this next question asks . . . "
4. Avoid directive reference to past responses.
   - As an interviewer, you must not "direct" the respondent toward an answer.
   - Do not assume that an "answer" you received in passing is the correct answer to a specific question at a particular point in the interview. Do not direct the respondent by mentioning an earlier answer.
   - If an answer is different from the one you expect, do not remind the respondent of an earlier remark or try to force consistency.

Style

1. The quality of your delivery—your *style*—also affects the quality of the information you collect. Emphasize underlined words to enhance meaning and to be consistent with the other interviewers. Keep your tone neutral, and avoid voice inflections that might bias results.
2. Use a pleasant tone of voice that conveys assurance, interest, and a professional manner.
3. Read at a slow pace. Remember that although you may have read these questions many times, the respondent is hearing them for the first time. He or she needs time to understand the questions and to decide on the answers.

Question Clarification

Sometimes respondents ask for additional information. When this happens, your response is called a CLARIFICATION. There are three kinds of clarifications:

1. Accurate repetition of the entire question, or part of the question.

2. Use of clarifications or definitions that are specified on the definitions sheet.
3. Use of the phrase, "Whatever________ [Insert the word or phrase about which the respondent asked.] means to you," or "Whatever you think of as ________."

Points to Remember

1. If you have any doubt that the respondent has heard the entire question, repeat *all* of it.
2. Upon request, repeat or clarify the item to which the respondent has referred.
3. When asked to repeat only one of several response options, repeat all the options given in the question.
4. Only give definitions specifically allowed in your interviewer's manual.

Examples of Standard Probes and Clarifications

1. Whatever ________ means to you.
2. Whatever you think of as ________.
3. What do you mean?
4. How do you mean?
5. Would you tell me more about your thinking on that?
6. Would you tell me what you have in mind?
7. What do you think?
8. What do you expect?
9. Which would be closer to the way you feel?
10. Are there are any other reasons why you feel that way?

Examples of Special Probes

1. Would that be *good* times or *bad* times?
2. Would that be *favorable* or *unfavorable?*

Examples of Neutral Ways to Preface a Probe

1. Overall . . .
2. Generally speaking . . .
3. Well, in general, . . .
4. In the country as a whole . . .
5. Yes, but . . .
6. Of course no one knows for sure . . .
7. Of course there are no right or wrong answers . . .
8. We all hope, but . . .
9. We're just interested in what you think . . .
10. Let me repeat the question. . . .

Pausing or Repeating a Question

Thoughtful answers are more complete and more accurate. Give the respondent time to think by reading the question slowly and pausing as appropriate. As you gain experience, you will learn how long a pause should be. Keep in mind that the respondent has never heard the questions before, so a pause that seems long to you might be just right for the respondent.

Don't rush your respondents—pausing is another way of indicating that you expect thoughtful answers.

1. Repeat the entire question if the respondent's reply indicates that they didn't understand it, or if they need more time to think about their response.
2. If the respondent has clearly eliminated a response option, you do not have to include it in the repetition.

3. Repeat the entire question unless you are sure that only one part of it was misunderstood.

Open-ended Responses

After listening to a respondent's answer to an open-ended question, repeat the response that you have written down. Make sure your words are acceptable to the respondent.

Ways to Probe for Answers on Open-ended Questions

1. When you cannot understand the respondent's reply:

   "What do you mean?" or "Could you tell me what you mean by that?"

   "Would you tell me more about your thinking on that?" or "Would you tell me what you have in mind?"

2. When the respondent gives an incomplete answer:

   "Would you tell me more about your thinking on that?" or "Would you tell me what you have in mind?"

3. When the respondent responds with a tentative "I don't know" or "I hope so":

   "What do you think?" or "What do you expect?"

4. When the respondent has narrowed the choices to two or to a range between two:

   "Which would be closer?" or "Which would be closer to the way you feel?"

5. When an open-ended question asks why:

   "Are there any other reasons you feel that way?" *Use only once!*

What is Feedback?

Feedback consists of statements or actions that indicate to the respondent that he or she is doing a good job.

1. Effective interviewers give feedback only for acceptable performance, not "good" content.
2. Give short feedback phrases for short, one- or two-word responses (usually for close-ended questions).
3. Longer, more thoughtful answers deserve longer feedback (usually for open-ended questions).
4. A brief pause followed by a feedback phrase makes the feedback more powerful.
5. Specific study information and interviewer task-related comments also act as feedback because they motivate the respondent.
6. Telephone interviewers should give feedback for acceptable respondent performance from 30 to 50 percent of the time.

Examples of Positive Feedback

1. Short
   - I see. . . .
   - Uh-huh/Um-hmm.
   - Uh-huh/Um-hmm, I see.
   - Thank you.
   - Thanks.
2. Long
   - That's useful/helpful information.
   - It's useful to get your ideas on this.
   - Thanks, it's important to get your opinion on that.
   - I see, that's helpful to know.

- It's important to find out what people think about this.
- That's useful for our research.

3. Interviewer task-related comments
   - Let me get that down.
   - I need to write it all down.
   - I want to make sure I have it right (REPEAT ANSWER).
   - We have touched on this before, but I need to ask every question in the order that it appears in the questionnaire.

Feedback phrases and neutral probes may be used in any combination. For more information, see the Survey Research Center's *General Interviewing Techniques,* cited in the reference list at the end of this book.

9

From Questionnaires to Survey Results

Stacks of completed questionnaires are now accumulating. Especially for people who have never done a survey before, the piles can be intimidating. Progress sometimes seems to grind to a halt as work enters a new phase. So this chapter is intended to lead you through the next step. As illustrated in Figure 9.1, the goal is to go from unprocessed questionnaires to results that people can understand and digest.

Briefly, this chapter explains how to

- design a coding system that assigns a number to every answer in the questionnaire;
- make a master list of all the codes;
- go through the questionnaires to edit answers and code those that aren't already in numerical form;
- enter the data from the questionnaires into a computer (unless you tabulate by hand);
- summarize and analyze the data; and
- interpret the results.

Before getting into the heart of the chapter and describing these tasks in detail, we take a short detour into the subject of computers. Specifically, we address whether to use a computer at this stage in the survey and, if so, what kind of software might work. These two questions need to be answered before you can begin getting the data ready for analysis.

Should You Use a Computer?

Questionnaires from some surveys can be tabulated by hand, without using a computer at all. For example, parents and teachers at an elementary school were recently surveyed about topics they'd like to hear discussed at the annual PTA forum. The PTA had received a grant to bring speakers to the school and decided to do a one-question survey. Its only purpose was to find out which subject would draw the biggest crowd. Hence, the survey form had just a single question:

> Which one of the following topics would you be MOST interested in hearing discussed at the upcoming PTA forum? (Please circle the number of your response.)
>
> 1 HELPING CHILDREN UNDERSTAND PEER PRESSURE
>
> 2 RESOLVING CONFLICTS BETWEEN CHILDREN

Figure 9.1
An example of turning unprocessed questionnaires into results people can understand

Q-16 Last month, residents of the North Central School District were asked to vote on a proposed 20-year school bond levy of $15.1 million to build a new high school and make other improvements in district facilities. Were you aware of this proposed bond levy?

1 YES
2 NO → SKIP TO QUESTION #20

Q-17 As you think back to that election, which of the following best describes how favorable or unfavorable you were towards the approval of the bond levy? Would you say that you were strongly favorable, somewhat favorable, somewhat unfavorable, or strongly unfavorable towards its approval?

1 STRONGLY FAVORABLE
2 SOMEWHAT FAVORABLE
3 SOMEWHAT UNFAVORABLE
4 STRONGLY UNFAVORABLE
5 DON'T REMEMBER OR DON'T KNOW

Q-18 Next, I have a few questions about you and your household. Which of the following best describes where your home is located?

1 SOUTHPORT NEIGHBORHOOD
2 CARRIAGE HILL NEIGHBORHOOD
3 COLLEGE NEIGHBORHOOD
4 PIONEER NEIGHBORHOOD
5 OUTSIDE CITY LIMITS

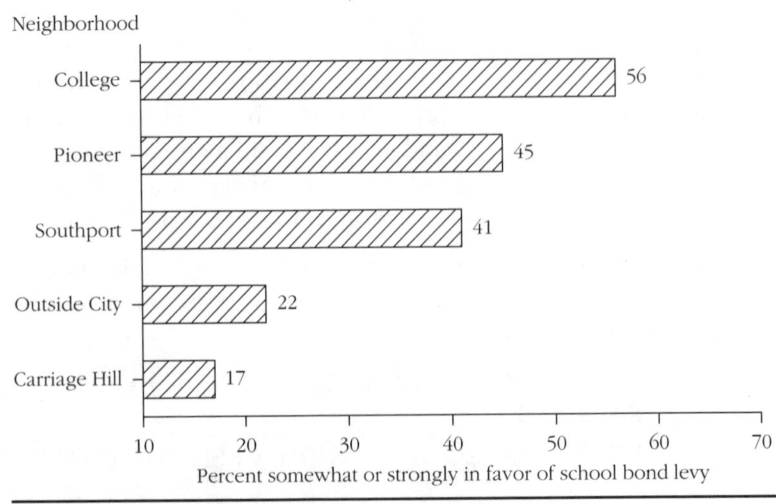

3 OFFERING ALTERNATIVES TO TV
4 AVOIDING GENDER STEREOTYPES
5 HELPING TEACHERS BY SUPPORTING INSTRUCTION AT HOME

A total of 204 questionnaires were returned. It took survey organizers only an hour to hand-tabulate the results and find out which topic was the most popular. All they had to do was count the number of responses for each item. Their results looked like this:

Topic	Number of responses
1. Helping children understand peer pressure	31
2. Resolving conflicts between children	43
3. Offering alternatives to TV	68
4. Avoiding gender stereotypes	20
5. Helping teachers by supporting instruction at home	42
Total	204

Single-question surveys like this one are very unusual. We've used the example to show that tabulating by hand does work in some cases. Most surveys are motivated by more complicated information needs that involve looking at two questions at the same time. For example, staff at a community action agency recently did a survey on child care. Their goal was to find out whether parents were satisfied with the day care that was available locally. Parents were asked a number of questions about their work schedule, ability to pay for day care, the age of their children, and problems with current arrangements. Two hundred questionnaires were mailed out and 78 percent were returned.

Even though only 156 completed questionnaires were involved, tabulating results for this survey was much more complicated than for the survey in our previous example. That's because analyzing the data meant comparing subgroups of parents and looking at how answers to certain questions were related to each other. For example, the organizers wanted to know whether parents who used home care providers had different problems than those who used day care centers. In other words, they had to use the answers to *two* questions at a time and needed a table like the one in Figure 9.2.

Figure 9.2
A example of how two questions can be analyzed at once

What working parents say about problems with child care (percentage who say problem is serious)

Problem	Type of care		
	Home care providers	Day care centers	Total
Cost is too high	43	31	37
Not always available when needed	32	18	25
Kids are exposed to illness	10	66	38
Inconvenient location	11	21	16
Too impersonal	4	36	20
High staff turnover	—	48	24

The survey organizers had many more questions as well. They wanted to know whether parents with younger children had more trouble finding adequate care than those with older children, how much of their income was spent on child care, and a host of other information.

For a survey like this, hand-tabulation would involve sorting questionnaires into piles, counting answers, and calculating results. Then questionnaires would have to be resorted, answers recounted, and results recalculated over and over—a time-consuming and inefficient way to get information. Instead, organizers of the day care survey used a microcomputer to analyze their data. They typed answers from each questionnaire into the machine, which in turn did the calculations.

The advantage of using a computer is that once the data are entered, one can do any number of counts and comparisons. Computers provide enormous flexibility in terms of which questions are analyzed and in what combination.

There are several options if computer-tabulation seems appropriate for a survey but you don't have access to a machine. For example, a statistics or social science instructor at the local community college might be interested in assigning his or her students to do the analysis as a class project. Or someone in the community might loan their computer if you provide the software.

For those who have never used a computer before, it is very likely that help will be required at this stage of the project. The skills involved are reasonably modest but not everyone enjoys this kind of challenge. Even those who do may want someone to help them get started.

What About Software?

As personal computers have grown in popularity, software has become more specialized and powerful. The sheer number of programs available from both small and large software companies is daunting. One can buy programs to organize conferences, compute taxes, keep payroll records, schedule work, track stock prices, balance checkbooks, draw maps, write reports, learn typing, and on and on.

We suggest that the less experience you've had with computers, the more important it is to find software that is simple and easy to use. The first thing to look for is a program that lets you both enter *and* analyze your data—tasks that we describe later in this chapter. You can get more power and sophistication by buying one program for data entry (a **database** program such as dBASE) and one for analysis (a **statistical** program such as SPSS-X). However, many people who use this book will not have extremely large data sets or need to do complicated statistical analyses and can therefore get by with software that combines data entry and analysis.

Another useful feature to look for in software is the ability to read in and write out data files that can be handled by other programs. In computer jargon, this is called *importing* and *exporting* **ASCII** (pronounced ASK-ey) *files*. You can think of these files as being written in a standard or generic language that other programs can understand. Having the ability to read and write ASCII files means you can take data entered using someone else's software (like a database program, for example) and analyze it on your computer. You can also write out data that can be read by other software (like a graphics or word-processing program). The ability to import and export ASCII files makes it much more likely that you'll be able to transfer your data between other programs as the need arises.

Finally, it is useful to find software that can display data in a graphic form, in other words, to make bar charts, pie charts, line graphs, plots, and other pictures. In Chapter 10, we'll talk more about the value of using good graphics when presenting results. For now, you should know that some data entry and analysis programs have a graphics feature. If you find one that does, you won't have to buy a separate piece of software to perform this function.

There are many programs that combine the features discussed above. An example of one we have used is Number Cruncher

Statistical System (NCSS). Written for IBM and IBM-compatible computers and marketed by a company in Keysville, Utah, NCSS is relatively inexpensive and easy to use.

Editing and Coding Survey Data

Now it's time to get the questionnaires ready for data entry and analysis. This is time-consuming work that needs to be done in every survey involving more than a few questions, especially when they are open-ended. The longer the survey, the more work is entailed.

Editing and coding are often done at the same time, but for the moment we'll treat them as separate tasks. The purpose of editing is to **clean** the questionnaires, which means several things. First, it means making sure that whoever enters the data into the computer will know which marks are actually answers and which are extraneous notes that someone might have made during the survey. This is especially important if you've arranged to have data entry done by someone who isn't familiar with the project.

Cleaning the questionnaires also means getting rid of obviously erroneous responses that make no sense at all, sometimes called **outliers**. Suppose that in response to a question about the year in which they were born, someone gave the year of the survey by mistake. If you left an answer like this alone and treated it as if it were valid, the final results would be very odd, not to mention inaccurate. (It is possible to let the computer identify outliers at a later stage but we'll assume the job is being done by hand.)

Finally, cleaning the questionnaires means checking to make sure that each one tells a consistent story. On one hand, this simply involves checking to see that skip patterns were followed correctly. On the other, it can require some knowledge of the subject matter and reasonable judgment. In a survey of teachers, for example, a respondent might say he has been employed by the school district for 20 years but also report an entry-level salary. In cases like this, you need to decide whether there is a logical explanation and the apparent inconsistency can be ignored, or if it needs to be dealt with. That can be as simple as correcting transposed numbers or as difficult as recontacting the respondent. Failing to examine and, in some cases, deal with inconsistencies increases the risk of measurement error.

Coding the questionnaires means expressing in terms of numbers all responses that will eventually be analyzed. For the vast majority of surveys, we recommend that people develop a master list or codebook in which to keep track of all the codes used in the survey, including the ones that appear on the questionnaire and those that are added after data collection. Some people simply make notations on a blank questionnaire; others use a binder or other notebook format.

The codebook is an essential reference for whoever enters data into a computer. Not only does it list and number all the variables, it also labels them and tells which values are legitimate.

A codebook is unnecessary only for short and simple surveys in which, first, no open-ended or partially close-ended questions are asked, and second, missing values don't present a problem. In that case, all answers are already expressed in numerical form and therefore can be tabulated without additional work, either by hand or computer.

In Chapter 6, we described four categories of question structure: close-ended with ordered choices, close-ended with unordered choices, open-ended, and partially close-ended. The first two involve one set of coding issues, while the third and fourth pose different problems. We use Figures 9.3 and 9.4 to show how these different kinds of questions are coded. The first illustrates an excerpt from a face-to-face questionnaire and the second shows part of the codebook that goes with it.

Close-ended Questions

Whether the responses are ordered or unordered, most close-ended questions can be written in such a way as to be **self-coded**. That means each possible response has a number printed next to it in the questionnaire. Question 2 in Figure 9.3 is an example.

Some close-ended questions, like Question 3 in Figure 9.3, are trickier to code because they give respondents the opportunity to select more than one answer. One way to handle such **multiple response** questions is to think of each answer choice as a *separate* variable. Instead of being associated with just one variable, Question 3 has five: "paid vacation," "paid sick leave," "pension plan," "health insurance," and "none of the above." In effect, the respondent answers yes or no to each. As Figure 9.4 shows, each variable gets coded either yes or no, depending on whether or not it is circled. If the first four are coded no, then the fifth should be coded yes (unless the respondent skipped the question altogether).

Open-ended Questions

Questions 5, 6, and 7 in Figure 9.3 are all open-ended questions, but each involves a different kind of coding. Question 5 asks respondents to provide a number, so it is self-coded just as if it were close-ended.

Figure 9.3
Excerpt from a face-to-face survey of local businesses

Q-1 We would like to begin by asking some questions about your business. Did you:

1 BUY AN EXISTING BUSINESS?
2 START YOUR OWN BUSINESS?
3 INHERIT YOUR BUSINESS?
4 OTHER ___________

Q-2 How many people do you employ in addition to yourself?

1 NO OTHER PEOPLE → If no employees, SKIP TO Q4
2 1–2 OTHER PEOPLE
3 3–5 OTHER PEOPLE
4 6–10 OTHER PEOPLE
5 MORE THAN 10 OTHER PEOPLE

Q-3 Which, if any, of the following employee benefits do you provide? **(Circle all that apply.)**

1 PAID VACATION
2 PAID SICK LEAVE
3 PENSION PLAN
4 HEALTH INSURANCE
5 NONE OF THE ABOVE

Q-4 How is your business organized for legal and tax purposes?

1 SOLE PROPRIETORSHIP
2 PARTNERSHIP
3 CORPORATION

Q-5 How many years ago did you buy, start, or inherit your business?

___________ years ago (If less than 12 months, enter "1".)

Q-6 What are the principal products or services provided by your business? What percent of gross sales or revenues is contributed by each?

PRODUCT OR SERVICE: PERCENT:

______________________ ____

______________________ ____

Office Use: ____

Q-7 What is the single most important reason you went into business for yourself?

____

____

How to Code Occupation and Industry

As mentioned several times in this book, coding answers to open-ended questions can be a real challenge. This is especially true for questions about what people do for a living (their occupations) and the kind of business they work for or operate (their industry). Luckily, the Federal government has developed standard coding schemes for occupations and industries. Both schemes are hierarchical, which means they go from very general to very specific.

At its most general level, the occupational scheme has 22 divisions:

Executive, administrative, managerial
Engineers, surveyors, and architects
Natural scientists, mathematicians
Social scientists, social workers, religious workers, and lawyers
Teachers, librarians, and counselors
Health diagnosing and treating practitioners
Registered nurses, pharmacists, dieticians, therapists, and physicians' assistants
Writers, artists, entertainers, and athletes
Health technologists and technicians
Technologists and technicians, except health
Marketing and sales
Administrative support, including clerical
Service
Agricultural, forestry, and fishing
Mechanics and repairers
Construction and extractive
Precision production
Production working
Transportation and material moving
Handlers, equipment cleaners, helpers, and laborers
Military
Miscellaneous

At its most general level, the industrial scheme has 10 divisions:

Agriculture, forestry, and fishing
Mining
Construction
Manufacturing
Transportation, communications, electric, gas, and sanitary services
Wholesale trade
Retail trade
Finance, insurance, and real estate
Services
Public administration
Nonclassifiable

We recommend that you use the federal coding schemes if you do a survey in which people are asked about what they do and where they work. Using existing schemes makes coding easier and also allows you to compare your survey results with secondary data sources such as the Census of Population and Housing.

See *Standard Occupational Classification Manual* and *Standard Industrial Classification Manual* for more detail. These are available in some libraries. If not in yours, contact the U.S. Government Printing Office at 202/783-3238.

Figure 9.4
Page 1 from a codebook developed for the questionnaire in Figure 9.3

General instructions:

- Enter only numerical codes, no text.
- Enter only questions listed below; others will not be analyzed.
- Code "No response" as 99, "Don't know" and "No opinion" **as 88.**

Specific instructions:

	Var 1	ID number from questionnaire front cover		
		Variable name	BUSID	
Q-1	Var 2	Variable name	ORIGIN	
		Value labels	1	BUY
			2	START
			3	INHERIT
Q-2	Var 3	Variable name	EMPLOYEES	
		Value labels	1	NONE
			2	1–2
			3	3 TO 5
			4	6 TO 10
			5	MORE THAN 10
Q-3	List of five answers; value label 1 if circled, 2 if not circled			
	Var 4	Variable name	PAIDVAC	
	Var 5	Variable name	PAIDSICK	
	Var 6	Variable name	PENSION	
	Var 7	Variable name	HEALTHINS	
	Var 8	Variable name	NOBENS	

Continued on next page

Question 6 in Figure 9.3 asks respondents to describe in words the product or service provided by their business. This information will enable coders to use a standard coding system that has been developed by someone else, in this case, a system for classifying kinds of businesses. Another standard system is designed to classify types of jobs. (See the shaded text How to code occupation and industry.)

The person who codes Question 6 will use the codebook almost like a dictionary. Using words or phrases provided by respondents, he or she will classify the business according to industry and, at the same time, translate respondents' answers into numerical form.

Of the three open-ended questions in Figure 9.3, Question 7 would probably take the longest to code. First, the person doing the

Figure 9.4
Continued

Q-4	Var 9	Variable name	BUSORG	
		Variable labels	1	SOLE
			2	PARTNER
			3	CORPORATION
Q-5	Var 10	Variable name	BUSYEAR	
Q-6	Var 11	Variable name	BUSTYPE	
		Variable labels	1	AG, FORESTRY, OTHER RESOURCE
			2	CONSTRUCTION
			3	MANUFACTURING
			4	TRANSPORTATION, UTILITIES
			5	TRADE
			6	FINANCE, INSURANCE, REAL ESTATE
			7	SERVICE (HEALTH, EDUCATION, OTHER)
			8	MISCELLANEOUS
Q-7	Var 8	Variable name	REASON	
		Variable labels	1	BUSINESS WAS IN THE FAMILY
			2	HAD AN IDEA FOR PRODUCT OR SERVICE
			3	GREATER INDEPENDENCE
			4	GREATER INCOME POTENTIAL
			5	UNABLE TO FIND WORK
			6	TIRED OF WORKING FOR SOMEONE ELSE
			7	OTHER ________________

coding must go through enough of the questionnaires to develop an adequate system that captures variation among answers but is still concise enough to be meaningful. Next, this person or someone else must go through all the questionnaires again to do the actual coding.

You may choose not to code some of the open-ended questions in your survey. For example, you might decide to ignore responses to an open-ended question that most people simply didn't answer. Or you may have included a few questions just to give people a chance to express their opinions rather than to provide information pertinent to your survey objectives.

Partially Close-ended

Question 1 in Figure 9.3 is a partially close-ended question because it allows respondents to pick from a list of answers or to formulate something in their own words. This format rarely yields many additional responses but has the advantage of not forcing respondents into predefined categories that aren't applicable to their situation. The best way to code partially close-ended questions is to make a list of responses during the editing process. Then an appropriate system can be developed if there are enough answers to warrant categories in addition to "other."

Missing Data

When data analysis begins, it is extremely important to be able to distinguish between "zero," "I don't know," and no response at all. For example, when calculating an average, you don't want to confuse "zero" with "I don't know." And it is likely that you will want to calculate the average based on the number of people who answered the question, not everyone in the sample (including those who inadvertently skipped the question). For this reason, we recommend reserving codes like 9 or 99 for "no response" and 8 or 88 for "don't know" or "no opinion." Such coding instructions are usually given at the beginning of the codebook, as shown in Figure 9.4.

Almost all computer software that you're likely to use in your analysis recognizes missing value codes as those that should be excluded from calculations.

Entering Data into the Computer and Double-Checking Your Work

The next step is to enter data from the questionnaires into a computer. The exact procedure will depend on the particular brand of software that you've chosen, but the basic task is usually the same: You go through each questionnaire, one at a time, typing in the responses in order.

To understand what happens when you enter numbers into a computer, think of the survey data as a series of rows and columns. Responses from each questionnaire make up one row of data, usually called a record. Each column contains everyone's response to a particular question. Figure 9.5 shows the data for the first six and the tenth variables in the first ten questionnaires in the local business survey that we have been using as an example. Row 1 contains data from the questionnaire filled out by the proprietor of Kate's Ice Cream Store, row 2 contains data from the proprietor of Micro Movies, and so on. Column 1 contains an identification number for each questionnaire

("BUSID"), column 2 contains data for the variable "ORIGIN," and so on. If there were 250 responses to this survey and 100 variables in the questionnaire, we would have 250 rows and 100 columns.

Most computer programs that are designed for data entry have a mode that allows you to enter one questionnaire's worth of data (or one record) at a time. Figure 9.6 shows what the data entry screen looks like in the program NCSS, and how the screen changes as one proceeds through a questionnaire, in this case, from BUSID to ORIGIN to EMPLOYEE. The variable number and name appear at the top of each screen. The shaded box tells you the computer is waiting for a specific value. After you have entered all the data from the first questionnaire, the program switches to row 2 and you can then enter data for the second questionnaire.

Note one very important point about ensuring respondents' confidentiality. As data from the questionnaires get entered into a computer, all identifying information is stripped away. Hence, Kate's Ice Cream Store becomes simply BUSID 1, Micro Movies becomes BUSID 2, and so on. The result is that no one using the data can associate information with a particular respondent.

Even people who are experienced in data entry make mistakes—they transpose digits, enter some responses twice, and skip others. For this reason, we suggest that after entering data from all the questionnaires, you repeat the whole procedure again and do some basic

Figure 9.5

Rows and columns of data for the first six, and the tenth, variables from questionnaires in a local business survey

Business Name[a]	BUSID	ORIGIN	EMPLOYEES	PAIDVAC	PAIDSIC	PENSION	BUSYEAR
Kate's Ice Cream	1	1	3	2	2	2	3
Micro Movies	2	1	2	2	2	2	7
Book Barn	3	3	3	2	1	1	12
Gems & Jewelry	4	2	3	1	1	2	2
Wally's Gas & Auto	5	1	2	2	2	2	22
The TV Shop	6	3	2	2	2	2	6
Eagle Construction	7	2	4	1	1	1	10
My Big Truck	8	2	1	99	99	99	11
Gold Bank	9	2	5	1	1	1	3
Energy Aerobics	10	2	1	2	2	2	1

[a]Identifying information like this may appear on the questionnaire. To ensure confidentiality, do *not* enter into the computer.

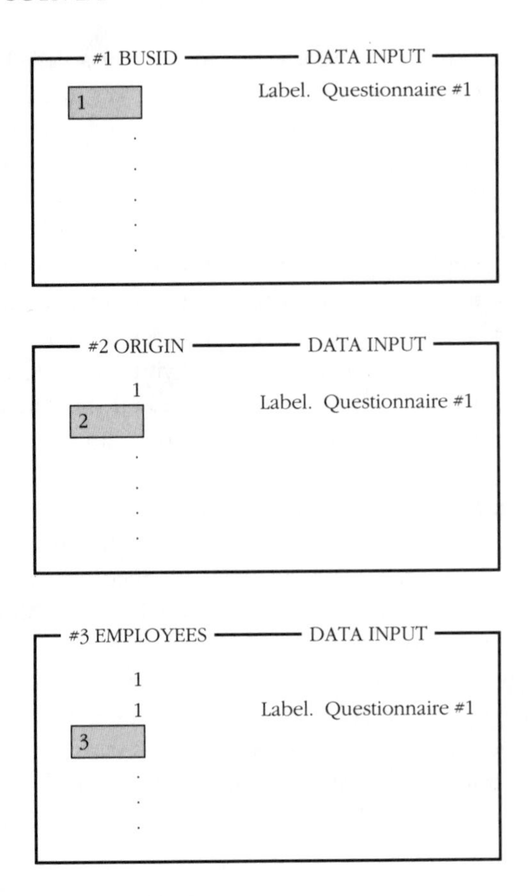

Figure 9.6
Three different computer screens as they appear during data entry for the first questionnaire in a local business survey

calculations on both data sets. By comparing results of calculations like those discussed in the following section, you can test whether the same numbers have been entered twice. This process is called **verifying** data. It is routinely done in professional surveys and is intended to avoid measurement error due to incorrect data entry.

Analysis

After data from all the questionnaires have been entered into a computer, it's time to start producing results. The purpose of this section is to help readers figure out which calculations make sense for the particular questions that have been posed. We introduce basic concepts so that descriptive statistics derived from a survey can be interpreted and used. The following pages are not meant to substitute for a thorough treatment of sample statistics, but we hope they will get readers started with an analysis.

Even if you want the very simplest summary numbers from a survey, you still have to decide between several different types of results. The following example from Figure 9.3 illustrates this point:

Q4 How is your business organized for legal and tax purposes?

1 SOLE PROPRIETORSHIP
2 PARTNERSHIP
3 CORPORATION

At first glance, it might seem reasonable to figure out the "average" response. However, adding up all the answers to this question and dividing by the number of respondents (which you would do to get a kind of average) would yield a number like "1.9." This wouldn't be very helpful. More informative would be a count and percentage distribution of how many people gave each response, as in Figure 9.7.

Not all questions lend themselves to percentage distributions. For Question 5 in Figure 9.3 ("How many years ago did you buy, start, or inherit your business?"), it would make more sense to calculate an average.

The problem is that many computer programs automatically provide percentage distributions, various kinds of averages, and dozens more statistics. The trick is to know which ones are most useful for you.

Looking at Answers to One Question at a Time

The first step in the analysis is to look at answers to individual questions, one at a time. We recommend doing this not only because the results themselves are important, but also because it is a convenient way to verify or check that data have been entered correctly.

In close-ended questions like the one about business organization, answers are already classified into groups. All you have to do is count—or have the computer count—how many people gave each response, and then calculate the right percentages.

Some close-ended questions have ordered responses where each choice represents a gradation of a single concept. An example is

Response	Number of respondents	Percentage of respondents
Sole proprietorship	110	44
Partnership	114	46
Corporation	26	10
Total	250	100

Figure 9.7
Showing answers to a close-ended question with a count and percentage distribution

Figure 9.8
Showing ordered responses to a close-ended question with a count and percentage distribution

Numbers of people employed by local businesses

Number of employees[a]	Number of responses	Percentage of all responses
None	75	30
1–2	58	23
3–5	52	21
6–10	43	17
More than 10	22	9
Total	250	100

[a] The respondent in this survey was the business owner. He or she is not counted as an employee.

Question 2 in Figure 9.3. Again, with this kind of question, it makes sense to count how many people gave each response, and then to calculate the appropriate percentages. The results might look like Figure 9.8.

Such a table enables you to make a statement like, "We estimate that 9 percent of local businesses have more than 10 employees." Moreover, it allows you to add up the responses from more than one category and say something like, "Similarly, 38 percent of the businesses surveyed reported between 3 and 10 employees."

Assuming you have already coded open-ended items like Question 6 in Figure 9.3, you can now summarize the results to these questions in tables similar to that in Figure 9.7. Again, you want to make some kind of statement about how many respondents fall into each of the categories you have delineated.

Other open-ended questions ask for a different kind of response. Question 5 in Figure 9.3 is an example: "How many years ago did you buy, start, or inherit your business?" Other examples:

- How many miles did you drive your car last year?
- What is your wage rate per hour?
- How many meals do you typically eat away from home each week?

Answers to questions like these are often summarized in terms of averages, or what are technically called **measures of central tendency**. Most computer programs *automatically* generate several kinds of central tendency measures, so you need to understand what each means. (See the shaded text Three Kinds of Central Tendency Measures.)

For the first ten questionnaires in the local business survey (see Figure 9.5), the mean value for BUSYEAR is 7.7, the median is between 6 and 7 (since there were an even number of responses), and the mode is 3.

Another way to summarize responses from open-ended items like Question 5 is to group the data and then make a count and percentage distribution table. In our previous examples, responses were *already* grouped, but now you must choose your own classification. For example, the 10 answers to Question 5 could be grouped in several ways, as shown in Figure 9.9. Option A, in which responses are most evenly distributed, is probably the most useful and Option C the least. However, there are no hard and fast rules about how to define the lower and upper limits on each group. We generally find it best to define categories so that each includes about the same number of responses. This usually takes some experimenting. How many groups we use depends on the need for precision—more can always be combined into fewer when results are presented.

So far, we have discussed counts, distributions, and measures of central tendency. There is one more useful calculation to consider: dispersion. It is often important to know how uniformly (or alternatively, how differently) respondents have answered a particular question. For example, in the business survey, we would likely want more than just the average number of years that firms had been operating. We would also want to know the distribution *across* firms. In other words, we are looking at the answers to find out if all businesses are

Three Kinds of Central Tendency Measures

Mean: sum of all responses divided by the number of responses

Median: middle value of all responses, when all responses are ranked from highest to lowest

Mode: response that has been given most frequently

Figure 9.9

Three ways to group answers to the question "How many years ago did you buy, start, or inherit your business?"

Option A		Option B		Option C	
Response	Number of responses	Response	Number of responses	Response	Number of responses
1–3	4	1–5	4	1–10	7
4–10	3	6–10	3	11–20	2
More than 10	3	11–15	2	More than 20	1
Total	10	16–20	0	Total	10
		More than 20	1		
		Total	10		

about the same age or if, instead, some are long established and others are very new. If we've asked for this information in an open-ended question, we can now calculate some measure of dispersion.

The simplest measure is the **range** from lowest to highest. For example, if the range in age of business is between 4 and 11 years, we have a very narrow range and little dispersion. If the range is 1 to 47 years, we have much greater dispersion, in other words, much less uniformity. The disadvantage of using a range to measure dispersion is that it captures only two extremes rather than the distribution of values in between.

A more useful measure of dispersion is the **standard deviation**. It measures how values are spread around a mean, in other words, whether they cluster together or are widely dispersed.*

Figure 9.10 shows three samples of 10 responses to Question 5 in the local business survey. All three samples have a mean of 7, but the first has a standard deviation of 0, the second of 1.76, and the third of 5.46. Hence, the third sample is the *most widely dispersed*. If we knew only the means, we would think the three samples of respondents were exactly alike. By knowing the standard deviations, we know they are very different.

The sample standard deviation is used to calculate a **confidence interval** around estimates derived from a sample (assuming a known distribution). The confidence interval is a range—centered around the estimate—within which we have a certain level of confidence that the true population value falls.

Consider the responses in Set B, Figure 9.10. The confidence interval is calculated around the mean 7 with standard deviation 1.76 and sample size 10. The **standard error** of the sample mean is the sample standard deviation divided by the square root of the sample size, in this case, 0.56. Knowing these numbers, we can make the following statement:

> Chances are 95 out of 100 that our estimate differs from the true population value by less than 1.1 in either direction. That is, we are 95 percent confident that the true mean is between 5.9 and 8.1 (7 minus 1.1 and 7 plus 1.1).

The size of the confidence interval depends on how large and how uniform the sample is. Larger samples and smaller standard deviations mean smaller confidence intervals. (See Appendix 5.A.)

---

*The sample standard deviation is one of the statistics customarily calculated by statistical software. It is the square root of the sum of the squared deviations from the mean divided by one less than the sample size.

Respondent #	Set A	Set B	Set C
1	7	7	3
2	7	4	6
3	7	9	10
4	7	4	2
5	7	8	19
6	7	7	6
7	7	7	9
8	7	8	11
9	7	7	3
10	7	9	1
Sample mean	7	7	7
Standard deviation of the sample mean	0	1.8	5.46
Standard error of the sample mean	0	0.56	1.73

Figure 9.10

Means, standard deviations, and standard errors for three different samples of 10 responses to Question 5 in the local business survey

This discussion may strike some of our readers as esoteric and of no importance whatsoever. But as we said in Chapter 2, sampling error is a fact of life for those who conduct sample surveys. By ignoring it, we risk interpreting the survey results incorrectly. We also open ourselves up to criticism from those who understand statistics and can hurt our credibility by pointing out errors in the analysis.

Looking at Answers to Two Questions at a Time

The discussion so far has focused on how to summarize results from individual questions for everyone in the sample. However, the value of many surveys lies not in a single statistic for the sample as a whole, but in comparable statistics for groups within the sample.

In the business survey we've been discussing in this chapter, an economic development group wanted to know the characteristics of local businesses that were expanding compared to those of stable or declining businesses. They were specifically interested in whether business size, which they measured by number of employees, was different between expanding and declining businesses. To get this information, they had to analyze responses to *two questions at a time*—first from Question 2 about number of employees (as a measure of size) and second, from one about changes in gross revenue that appeared later in the questionnaire.

Figure 9.11

Three ways to make a cross-tab showing the relationship between two variables, number of employees, and change in gross revenue

Panel A. Absolute numbers

	Change in gross revenue			
Number of employees	Increase	No change	Decrease	Total
None	31	16	9	56
1–2	24	14	11	49
3–5	19	17	20	57
6–10	11	14	15	39
More than 10	10	14	25	49
Total	95	75	80	250

Panel B. Column percentages

	Change in gross revenue			
Number of employees	Increase	No change	Decrease	Total
Total	95	75	80	250
None	33%	21%	11%	22%
1–2	25	19	14	20
3–5	20	23	25	23
6–10	12	19	19	16
More than 10	11	19	31	20
Total	101[a]	100	100	100

Panel C. Row percentages

	Change in gross revenue			
Number of employees	Increase	No change	Decrease	Total
Total	95	75	80	250
None	55%	29%	16%	100%
1–2	49	29	22	100
3–5	33	30	35	98[a]
6–10	28	36	38	102[a]
More than 10	20	29	51	100
Total	38	30	32	100

[a]Because of rounding, percentages don't always add up to an even 100.

One way of looking at two subgroups is to develop two-way tables, also called **cross-tabulations**, or *cross-tabs,* for short. The economic development group cross-tabulated responses from Question 2 and the gross revenue question. As shown in Figure 9.11, the results could be summarized in three ways, in terms of

- absolute numbers (Panel A);
- column percentages (Panel B); or
- row percentages (Panel C).

A cross-tab with absolute numbers is useful for making sure there are enough observations in each subgroup to work with, in this case, for businesses with increasing, stable, or decreasing gross revenue. However, without calculating percentages like those in Panels B and C, it is difficult to see any differences.

In Panel B, the percentages in each column add up to 100 (or close to 100 because of rounding). By comparing the columns, we can see that firms with increasing revenue tend to be smaller than those whose revenues are falling. For example, 33 percent of growing firms are very small and have no employees at all, compared to only 11 percent of firms whose revenues are decreasing. Hence, just by eyeing the table, we can see an apparent relationship between size and change in business activity as measured by gross revenue.

In Panel C, the percentages in each row add up to 100 (or close to it). By comparing rows, we can see that more firms with no employees are growing than are shrinking. In contrast, more firms with more than 10 employees are shrinking than are growing. This is the same information as presented in Panel B, only from a different perspective.

Tables like these are easily produced by computer programs designed for statistical analysis. The user simply has to specify which variable should appear on the rows and which on the columns. Such programs also routinely calculate a statistic called **chi-square,** a more rigorous and formal way of measuring whether a relationship exists between the row and column variables. (Refer to the documentation that comes with your software for information on how to use the chi-square statistic.)

Sometimes one or both of the variables that one wants to cross-tabulate come from open-ended questions, as in, "How many years ago did you buy, start, or inherit your business?" Since there are likely to be dozens of different responses to questions like these, cross-tabs rapidly become unwieldy. To avoid this problem, it is important to group or "recode" the variables. The process is easily done with computer programs and is the same one recommended earlier for developing count and percentage tables.

Another useful and more precise way to compare groups within a sample is to look at the **difference between means or proportions.** For example, in the survey discussed above, male- and female-owned businesses might be compared in terms of mean business age or the proportion that offer health insurance.

In comparing estimated means and proportions based on sample observations, we test whether differences are **statistically significant**, that is, large relative to sampling error. Sometimes means and proportions *appear* different, but because they are only estimates, we can't be sure without testing. Larger samples yield more precise estimates and therefore make true differences more likely to be detected. (See the shaded text on pages 197 and 198.)

Interpretation

At the end of the analysis phase of a survey, we usually have a large stack of computer printouts. While it takes up much less space than the piles of questionnaires with which we started, it often seems just as intimidating. How exactly does one interpret all those tables and statistics?

As we've noted several times already in this book, there is no foolproof recipe. Instead, here are some common sense suggestions to help structure the way you interpret the results. Taken together, these suggestions should help sort out which results are most useful and, at the same time, defensible.

Look for Results That Matter

Even for experienced researchers doing a survey of moderate length, the volume of results that comes out of the analysis phase can be overwhelming. Each question can be examined alone or in combination with any one of many others.

To sort the wheat from the chaff, it is usually best to start with questions that originally motivated the survey. What did you *need to know* in order to do your work better? After looking for this very specific information, ask yourself what else turned up in the survey that could potentially inform your work. Undoubtedly, you will have results that are *nice to know* as well—interesting tidbits you may or may not have expected. But start with actionable results that are pertinent to your work and file the rest away for later.

It is very important to look for the really BIG results, not the subtleties that don't matter much. Like so much of survey research, figuring out which are meaningful results involves making judgment calls based on personal knowledge of the situation. In the local business survey, organizers wanted to know which businesses were expanding and which were stable or declining. Their reasoning was

How Can One Be Confident That Two Sample Means Are Actually Different?

Sampling theory provides a formula for testing the difference between two means. If we want to test the difference between the mean ages of male and female-owned businesses and have a relatively large sample size, we would calculate

$$z = \frac{\bar{x}_1 - \bar{x}_2}{\sqrt{\frac{s_1^2}{n_1} + \frac{s_2^2}{n_2}}}$$

where z = test statistic

$\bar{x}_1$ = mean age of male-owned businesses

$\bar{x}_2$ = mean age of female-owned businesses

s_1 = standard deviation of x_1

s_2 = standard deviation of x_2

n_1 = number of male-owned businesses in the sample

n_2 = number of female-owned businesses in the sample

If $\bar{x}_1 = 10.2$, $\bar{x}_2 = 7.9$, $s_1 = 2.8$, $s_2 = 4.3$, $n_1 = 180$, and $n_2 = 123$, then

$$z = \frac{10.2 - 7.9}{\sqrt{\frac{(2.8)^2}{180} + \frac{(4.3)^2}{123}}}$$

$$= \frac{2.3}{\sqrt{\frac{7.84}{180} + \frac{18.5}{123}}}$$

$$= \frac{2.3}{\sqrt{0.04 + 0.15}}$$

$$= 5.2$$

For samples of this size, the difference is significant if the test statistic is greater than 1.96 (which we get from the standard normal (z) table in a statistics textbook). Here, we infer the two true means *are* different. (The significance level is 0.05.)

For more information, see *Statistical Analysis: An Interdisciplinary Introduction to Univariate and Multivariate Methods* by Sam K. Kachigan.

that they could get the biggest payoff by targeting a new financial credit program to the group of businesses that was already expanding. In this context, a difference of a few percentage points probably wouldn't be enough on which to justify a targeted development effort, but a difference of, for example, 20 percent could be very important.

Earlier in this chapter, we emphasized the value of testing for significant relationships between variables in a cross-tab and differences between means and percentages. Doing these formal, statistical tests

How Can One Be Confident That Two Sample Proportions Are Actually Different?

As is the case for sample means, sampling theory provides a formula for testing the difference between two proportions. If we want to test the difference between the proportions of male- and female-owned businesses that offer health insurance and have relatively large samples, we would calculate:

$$z = \frac{\hat{p}_1 - \hat{p}_2}{\sqrt{\bar{p}(1 - \bar{p})\left(\frac{1}{n_1} + \frac{1}{n_2}\right)}}$$

$$\hat{p}_1 = \frac{y_1}{n_1} \qquad \hat{p}_2 = \frac{y_2}{n_2}$$

$$\bar{p} = \frac{(y_1 + y_2)}{(n_1 + n_2)}$$

where z = test statistic

y_1 = the number of male-owned businesses that offer health insurance

y_2 = the number of female-owned businesses that offer health insurance

n_1 = the total number of male-owned businesses in the sample

n_2 = the total number of female-owned businesses in the sample

If $y_1 = 117$, $y_2 = 68$, $n_1 = 180$, and $n_2 = 123$, then

$$z = \frac{0.65 - 0.55}{\sqrt{(0.61)(0.39)(0.0137)}} = \frac{0.10}{0.057} = 1.75$$

For samples of this size, the difference is significant if z is greater than 1.96 (which we get from a table in a statistics textbook). Here, we infer the two population proportions are *not* different. (The significance level is 0.05.)

For more information, see *Statistical Analysis: An Interdisciplinary Introduction to Univariate and Multivariate Methods* by Sam K. Kachigan.

will strengthen your credibility and help defend your results if the need arises. When we say, "Look for big results," we are going one step further. The key point is that you must use your judgment to figure out which results are important enough to warrant action.

Be Careful When Working with Small Groups within the Sample

In Chapter 5, we recommended that you select a sample large enough to accommodate the analysis of subgroups. For example, if one of the questions motivating a survey of local businesses is whether small and large enterprises need different financial services, then it's important to have enough of each size business in the sample.

Even if you've been very careful to get a large enough sample, it is almost inevitable that some of the analysis will generate tables or counts in which the numbers are very small. For example, the result of counting businesses by number of employees might look like the following:

Number of employees	Number of businesses
None	107
1–3	115
4–10	47
More than 10	9
Total	278

Because of possible sampling error, the number of observations in the "more than 10" size classification is not enough to work with. One can't make estimates about the population of all businesses with more than 10 employees based on this small sample of 9. If one wanted to look at the characteristics of relatively large businesses, it would be better to break the classes differently, perhaps at "more than 7 employees," and then see if the number of observations is large enough to work with. As the number goes up, the sampling error goes down.

Summary

In the Preface we promised to keep this book as jargon-free and nontechnical as possible. Yet in this chapter, we discussed terms like *confidence interval, standard deviation,* and *test statistic*. We used these terms because they convey precise meanings and have no ready substitutes. In practice, they allow us to precisely calculate the likelihood that differences between groups are due to chance rather than real variation in a particular population. In other words, these concepts let us account for sampling error.

Still, we don't want to overemphasize the extent to which we can estimate survey error. Sometimes statistical tests and significance levels are taken as the true and final arbiters of the confidence we should have in our results. Once again, it is important to remember that the measures discussed here tell us nothing about coverage, nonresponse, and measurement error. Each of these must be assessed independently.

For example, it is important to think about the probability that your sample misrepresents the population in which you are interested. Were some members excluded from the sample because they had no chance of being selected? And further, were there many nonrespondents and do you have reason to believe they might be different in some important way from respondents? If the answer to either of these questions is yes, consider how the results have been affected.

Finally, think about errors that might have been introduced during the survey itself. When editing the questionnaires, did you notice any questions that didn't seem to work? Did people interpret certain items differently? Is there reason to believe that any of the interviewers biased respondents on particular questions? If the answer to any of these is yes, you may have to ignore certain results because of measurement error.

For more detail on coding, processing, and analyzing data, see:

Chapter 14, "Coding and Data Reduction" and Chapter 16, "Analysis, Presentation, and Interpretation of Data," in *Methods of Social Research,* Third Edition, by Kenneth D. Bailey. Macmillan, New York, 1987.

Chapter 11, "Data Processing" and Chapter 14, "Constructing and Understanding Tables," in *Survey Research Methods,* Second Edition, by Earl Babbie. Wadsworth, Belmont, CA, 1990.

Chapter 9, "Hypothesis Testing," in *Statistical Analysis: An Interdisciplinary Introduction to Univariate and Multivariate Methods,* by Sam K. Kachigan, Radius Press, New York, 1986.

10

Reporting Survey Results

The data have been collected, all the numbers have been analyzed, and now just one critical task remains. To make sure the new information gets used, the results must be presented logically and clearly. This step should be the highlight of the whole survey process, but it doesn't always work out that way. The brightest hopes are quickly dashed when people report results in a way that would-be users can't understand.

In one case, for example, a team that had surveyed community residents about satisfaction with public services waited to give their report until a "more complete" analysis of the data was finished. By delaying the presentation past its scheduled delivery, the surveyors heightened people's anticipation. At last, the long-awaited night arrived. The auditorium was packed with local residents, government officials, and news reporters.

A university professor addressed the audience, using an overhead projector and transparencies to show tables pulled directly from his written report. One by one and in the order they appeared in the questionnaire, he went through the answers to each of 20 questions dealing with attitudes toward public services. For example, he used a transparency to show how people answered the following question:

How satisfied are you with winter road maintenance?

1 COMPLETELY SATISFIED
2 MOSTLY SATISFIED
3 NEITHER SATISFIED NOR DISSATISFIED
4 MOSTLY DISSATISFIED
5 COMPLETELY DISSATISFIED

The transparency showed that the average response was 2.9. On he marched, through all 20 questions.

Next he explained how he had calculated a mean score and overall range for all public services. This, too, he showed with 20 sets of numbers on a transparency. At last, he presented the coup de grace. According to his regression analysis, 23 percent of the variation in people's satisfaction with public services could be explained by certain "independent variables." These included age, education, a dummy variable for home ownership, and last, distance from the county seat.

It was hard to say who was most stunned—the red-faced officials who sponsored the survey, a reporter whose notepad was blank, residents who donated time to the project, or the speaker who tried valiantly to explain the assumptions of regression analysis. Afterward the reporter remarked, "If I could have gotten the transparencies and read the print that was too small to see during the meeting, I might have understood why people were 3.7 satisfied on a five-point scale. But how could I tell my readers that 23 percent of their variation is explainable and the other 77 percent isn't?!?!"

A presentation like this might have been acceptable for academics interested in measuring and accounting for satisfaction with public services. However, it was completely wrong for people who didn't understand complicated statistics and simply wanted to know which services needed upgrading. It simply didn't meet the needs of the audience, and worse, it made people feel frustrated and confused.

A bad pitch isn't the only reason why some presentations fail. In another case, we watched survey sponsors walk into a room, proudly carrying stacks of printed tables and thick reports. Dropping their burden on the table, they announced, "Here are the results and we'd be happy to answer questions when you're ready!" Unfortunately, they assumed they could communicate their findings in writing without verbally summarizing the highlights.

In another instance, the presenter discussed responses to each of 85 questions in a survey. People in the audience looked dazed after the first 20 slides and soon began staring at the clock. What they really wanted was someone to translate all those numbers into usable information, someone to provide a summary that would help them make important decisions. Instead, all they heard was a boring recital about individual questionnaire items.

Finally, we've attended well-organized presentations pitched at the right level, but still received with confusion and frustration. On one memorable occasion, the presenter used colorful, eye-catching slides but constantly shifted back and forth between bar graphs, line graphs, and pie charts. The marvelous range and capability of her new computer graphics program served only to distract attention from the survey results.

Presenting survey results is not something to be taken lightly. This chapter describes the role and content of written and verbal reports, gives pointers on discussing errors, and then offers guidelines on how to use graphics effectively.

Written and Verbal Reports: Each Serves a Purpose

We recommend that people plan to report the results of their survey in writing as well as verbally. Each format compensates for deficiencies in the other because each is suited to a different audience and provides a different level of detail.

It is critical to understand that what makes an effective written report usually makes a dreary and boring verbal presentation, and vice

Major Sections to Include When Writing Up a Survey

- *Abstract or executive summary*—communicates the most important findings to hurried readers and gets people interested in opening the report.
- *Problem statement*—explains at the outset of the report why the survey was done and provides background on survey area (including map) if needed.
- *Methods and procedures*—describes survey method, sample design, period of study, and other details.
- *Error structure* (with respect to coverage, overall nonresponse, and sampling)—allows readers to evaluate how much and what kind of error might have crept into the results; greatly increases credibility by showing an awareness of possible shortcomings.
- *Findings*—presents survey results *that really matter* in logical order, not necessarily in the sequence that items appeared in the questionnaire; discusses item nonresponse, measurement, and sampling error where appropriate.
- *Implications*—draws the findings together to answer original questions and explores implications for decision making and future action.
- *Appendices*—provide supplementary material for the most interested readers, usually the questionnaire, and a listing or count of all answers for each question, including no opinion, refusals, etc.

versa. Written reports usually tell a more complete story. They can include particulars about the survey design, background on the study area, complex (but clear) tables of results, and a detailed discussion of what the findings mean. Readers are free to read the parts that interest them and skip the rest, taking as much time as they need to study the report.

Because it tells the whole story of a survey, one written report can meet the information needs of several audiences—a county commissioner, a summer intern, or a graduate student who wants to replicate the survey five years from now.

Completeness serves another purpose as well: It helps establish credibility. People who wonder whether the survey results are accurate can read a written report to learn how organizers dealt with potential sources of error.

In contrast, the person who makes a verbal presentation leads listeners through only the most relevant and interesting findings, typically in a very short period of time. Having sifted through all the results in advance, the presenter can pull out critical pieces of information and leave the rest for a written report. In addition, he or she can tailor the presentation to different audiences—one version for state legislators, another for local residents, and so on, depending on the nature of the study. Listeners can ask the presenter questions and refer to the written report for more information.

Using Visuals in a Verbal Presentation

We recommend using visual aids—either slides or transparencies—in a verbal presentation. In addition to helping the audience grasp key points, visuals work like note cards as we work through the presentation. They remind us of the most important points that need to be discussed. Our familiarity with the study allows us to add whatever details are appropriate for our audience.

To begin a verbal presentation based on survey results, we usually use the following series of visuals:

#1 A *title visual,* to introduce our study to the audience, as well as to make sure the overhead or slide projector is working and positioned correctly.

#2 A *brief statement of the objectives* to help establish people's expectations about what will be covered in the presentation. This is especially important if only some of the data or original project objectives are to be discussed.

#3 *Key points about the survey,* for example, when and how it was done, the sample size, and nature of the questionnaire.

#4–6 *Information about the accuracy of our survey estimates,* for example, specifics on the response rate, key characteristics of the respondents, and sampling error.

#7 *Exact wording and format of the first question* that will be discussed in the presentation, to help the audience understand potential measurement error.

#8 A *simple graphic* (often a pie chart) that displays results obtained from the question presented in the previous slide.

In subsequent visuals, we use slightly more complicated graphics (usually bar charts) to present objectives and results that are of greatest interest and value to the audience. In the last few visuals, we summarize the study's main findings, suggest implications for future action, and encourage our audience to discuss the survey results.

A friend who is devoted to using written reports once watched us make a verbal presentation with a complete set of slides. Afterward, he observed wryly, "It isn't that people who use slides know more than anyone else, they're just better organized and easier to understand!" Unintentionally, he had summarized the most compelling reason for using visuals in a verbal presentation: They help both the listener *and* the presenter.

It usually works best to develop a written report first and then prepare verbal presentations as the need arises. Of course, this isn't possible when people who want the new information can't wait for a written report. In general, however, putting the whole story on paper helps organize thinking and prods one to articulate study objectives, procedure, error structure, findings, and implications. After writing a report, it is usually easier to select salient points for verbal presentations.

Two pieces of advice: First, concentrate only on useful information and avoid getting bogged down in insignificant results. This is critical in verbal presentations but also valid in written reports. Second, it usually isn't necessary to report results in terms of decimal points. Just because computers calculate an average age like 34.792 or an income figure like $29,439.23 doesn't mean you are justified in claiming such accuracy. In any case, eliminating the decimals or at least rounding to one place usually helps people focus on the major points.

Reporting the Error Structure

By the time you actually have results to present, it can be very tempting to ignore the fact that your estimates aren't perfectly accurate. After all the work, it is hard to admit you can't present findings with absolute certainty, but of course it's true. At their very best, sample surveys only produce close estimates of what people think or do. By making this clear to the audience, you increase credibility and help keep people from misinterpreting the results.

In both written and verbal presentations, we usually discuss coverage, overall nonresponse, and sampling error at the outset. Then later, when we present the findings, we explain item nonresponse and measurement errors, and give more details on sampling error.

We begin by explaining whether the results are subject to coverage error, in other words, whether the frame matched the population of interest. For a telephone survey in which we sampled from a directory, we would explain that people without phones and those with unlisted numbers had no chance of being selected. For a mail survey of PTA members, we would explain that people who recently joined the organization might not have appeared on the membership list and so may not have been included in the sample. In both cases, we would try to figure out the extent of coverage error—whether it was unlikely to have an effect or, alternatively, might have biased the results in one direction or another.

Next we present the survey's overall response rate, explaining how many people declined to participate or simply could not be reached. Again, we tell people what is known about nonrespondents and therefore whether nonresponse error might be a problem.

Last, we talk briefly about sampling error. We explain that the findings are only estimates of actual population characteristics because we've surveyed a sample rather than conducting a census. We give the possible sampling error for estimates based on the sample and leave details until presentation of individual results.

Laying out the possibility of coverage, nonresponse, and sampling errors helps make clear to people the extent to which they can generalize from the sample. Remember that most of the audience probably won't distinguish between a sample and the whole population unless the presenter does it for them. They don't have the necessary experience in developing a list, drawing a sample, and calculating confidence intervals. Hence, it is the presenter's responsibility to make sure the audience interprets the survey findings correctly.

When it comes to presenting individual findings, we point out results that are especially sensitive to error for one reason or another. To help people understand the potential for error, we give the exact wording of questions about attitudes and beliefs. In a verbal presentation, we show the questions on a slide or transparency and, in a written report, we include the entire questionnaire as an appendix.

Using Graphics to Communicate Results

Bar charts, histograms, pie charts, and other graphics are powerful communication tools. They offer an excellent way to condense large amounts of numerical survey data and emphasize important findings. The key words here are *condense* and *important*. Using graphics to communicate clearly in a limited amount of time or space requires that we simplify to a great extent, especially in a verbal presentation.

An Example

One of the key questions motivating a recent needs assessment survey concerned public services. A group of county commissioners sponsored the survey. They wanted to know which services local residents rated highest and which they rated lowest. The questionnaire included a series of nine questions, each one about a different service.

Figure 10.1 shows how the commissioners' staff person first tried to summarize and compare answers to the nine questions. After carefully studying this table, the commissioners eventually would have reached the conclusion that residents rated road maintenance the highest and public health the lowest. Still, the table was more complex than it needed to be because all the columns tended to obscure rather than highlight key findings.

In Figure 10.2, the staff person condensed and reorganized the numbers. She rounded the percentages (from 6.2 to 6, for example) and focused only on services that were rated "fair" or "poor," so there were fewer numbers for the commissioners to absorb. She also eliminated the "no opinions" and recalculated the percentages. In doing so, she limited the table to people who actually responded to each item. And finally, she reordered the rows. Instead of listing services in the order they appeared in the questionnaire, she ranked them from lowest to highest quality.

How residents rate the quality of local services (n = 370)					
Service	**Poor**	**Fair**	**Good**	**Excellent**	**No opinion**
Police protection	6.2%	23.2%	28.2%	33.6%	8.8%
Education	11.0	20.9	22.1	32.4	13.6
Public health services	31.5	21.7	14.9	10.0	21.9
Building inspection	11.6	33.7	22.6	12.3	19.8
Fire protection	6.9	15.9	20.3	53.4	3.5
Road maintenance	5.4	13.4	18.5	58.7	4.0
Library	15.1	29.7	32.9	24.3	2.0
Public transportation	7.5	19.9	25.6	41.4	5.6
Garbage collection	22.3	22.6	18.1	22.2	15.0

Figure 10.1
The first draft: a comprehensive table that obscures important findings and is hard to read

Next she illustrated the results with a graph, simplifying them further by combining the percentage of respondents who selected "fair" or "poor" for each service (Figure 10.3). The commissioners could read the graph from top to bottom, quickly getting an answer to their original question, that is, how residents rate local services.

By highlighting the most important findings, the person who created the graph in Figure 10.3 lost some information contained in her original table. Figure 10.4 shows how a little of that can be worked

Local services with a rating of fair or poor (n = 370)			
Service	**Fair**	**Poor**	**Fair or poor**
Public health services	28%	40%	68%
Garbage collection	27	32	59
Building inspection	42	14	56
Library	26	16	42
Education	24	13	37
Police protection	25	7	32
Public transportation	21	8	29
Fire protection	16	8	24
Road maintenance	14	6	20

Figure 10.2
The second draft: a table in which results are condensed and reorganized, making it easier to grasp results

back in with slightly more sophisticated graphics. Still, the constraint we mentioned earlier is unavoidable. Getting the most out of graphics means condensing information. Too much information usually defeats the purpose of using a graph in the first place.

What's Involved?

The first step in graphically summarizing survey results is to articulate which question the graphic is supposed to answer. Sometimes, this question concerns just a single item from the questionnaire. More often, it has to do with one of the broader issues that motivated the survey in the first place. In our example above, the commissioners didn't care much about how people rated each individual service, but rather, how the ratings compared with each other.

The second step is to identify which variables you want the graph to illustrate and whether they should be expressed in terms of raw numbers, percentages, or both. Knowing which and how many variables are of interest and how they are expressed will help determine the kind of graphic that will work best.

Most computer graphics programs offer a wide range of options, including histograms and bar, pie, and line charts. The ones we use

Figure 10.3

The third draft: a simple graphic that reduces and highlights important findings

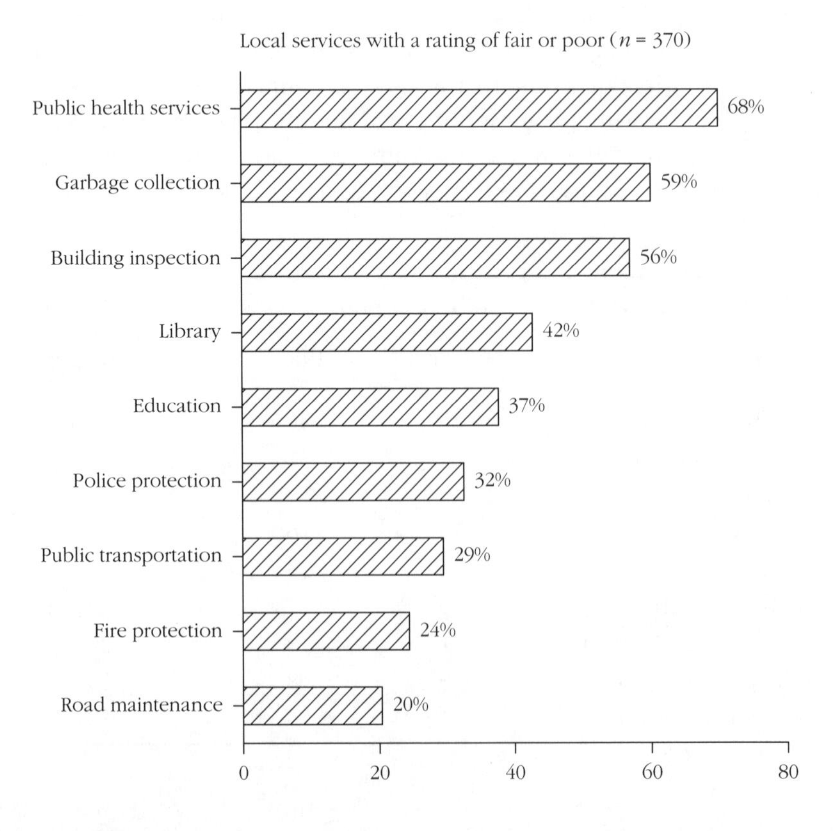

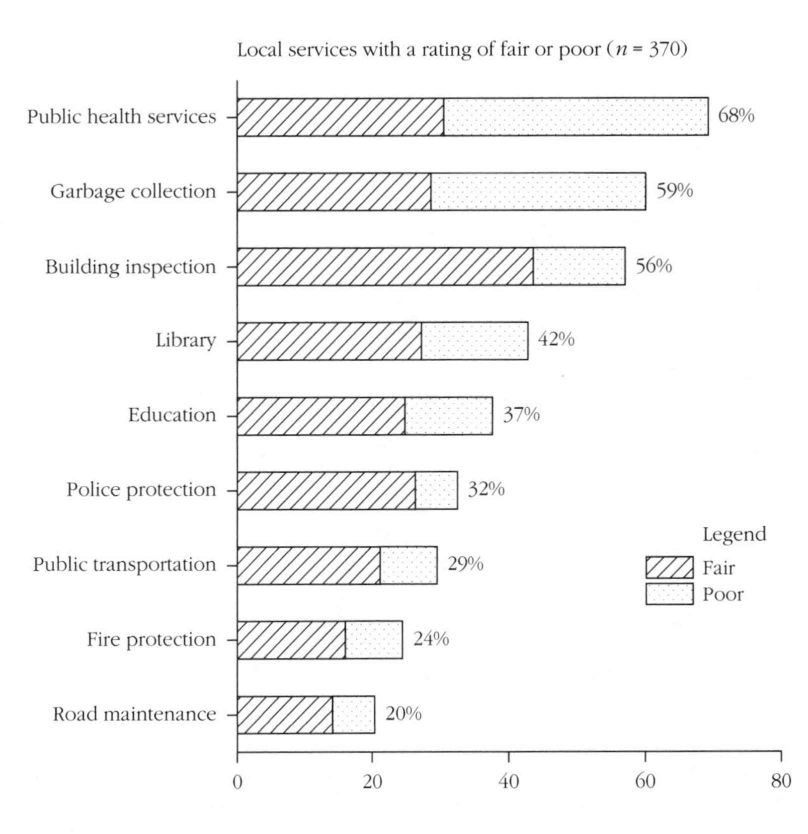

Figure 10.4
The final draft: a slightly more sophisticated graphic that highlights important findings and retains more detail

most often for illustrating survey results are bar and pie charts. There are no hard and fast rules for determining when one or the other is most appropriate. However, pie charts are generally best for showing the components of something that adds up to 100 percent, and bar charts are best for showing relative magnitudes.

For example, the pie chart in Figure 10.5 is used to show the percentage of respondents who selected each answer choice in the following question:

How many people do you employ in addition to yourself?

1 NO OTHER PEOPLE
2 1–2 OTHER PEOPLE
3 3–5 OTHER PEOPLE
4 6–10 OTHER PEOPLE
5 MORE THAN 10 OTHER PEOPLE

As a general rule, pie charts become harder to read as the number of categories increases. Bar charts work better if there are many categories to compare, and to show relative magnitude, as in Figure 10.4.

Figure 10.5
Using a pie chart to show how respondents answered one particular question

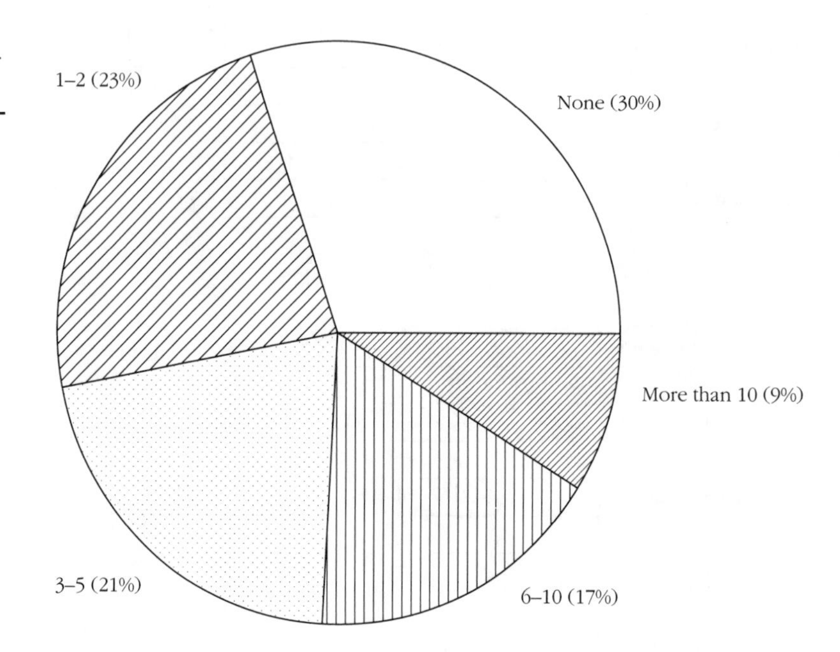

As noted in Chapter 9, the value of most surveys lies in comparable numbers for groups within a population. That means we often want a graphic to depict two different variables at the same time, in other words, to show numbers from a cross-tab. Earlier in the book we used data from a survey of local businesses to create Figure 10.6. The numbers in the table are graphically illustrated with a bar graph

Figure 10.6
Using a table to look at two variables at the same time

Local businesses by number of employees and change in gross revenue, 1990–93 (n = 250)

Number of employees	Change in gross revenue: Increase	No change	Decrease	Total
None	33%	21%	11%	22%
1–2	25	19	14	20
3–5	20	23	25	23
6–10	12	19	19	16
More than 10	11	19	31	20
Total	101	100	100	100

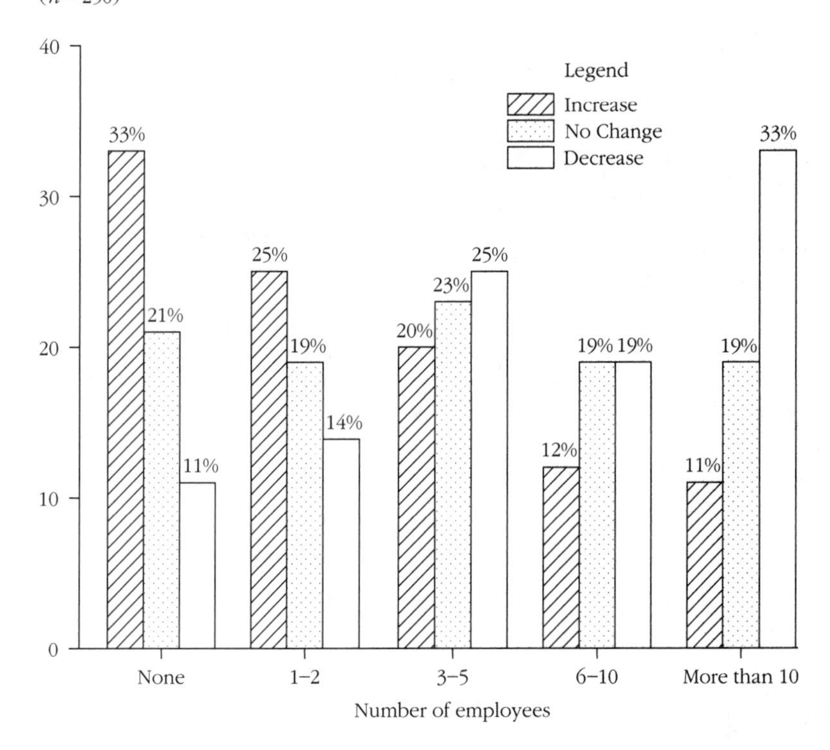

Figure 10.7
Using a bar chart to show how respondents answered two questions

in Figure 10.7. Each cluster of bars represents one row in the table. Note that this bar graph could have been constructed so that each cluster represented one *column* of numbers.

As discussed above, pie charts are useful for showing the components of something that add up to 100 percent—the occupational makeup of a labor force for example, or the percent of people with different levels of education. In contrast, **stacked** bar charts work well for showing components of something that is not expressed in terms of percentages. An example is illustrated in Figure 10.8. Here, data from a household survey are used to show the dollar value of income from different sources, by age of household head. Each column sums to the average income level for that group, $38,130 for households in which the head is under age 25 and so on. The individual components of household income are depicted by stacks within each bar.

Graphics in Written and Verbal Reports

Both written and verbal reports usually include graphics but rarely in the same form. In a written report, text makes up the centerpiece

Figure 10.8
Using stacked bar charts to show parts of a whole

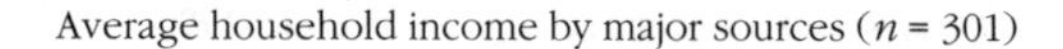

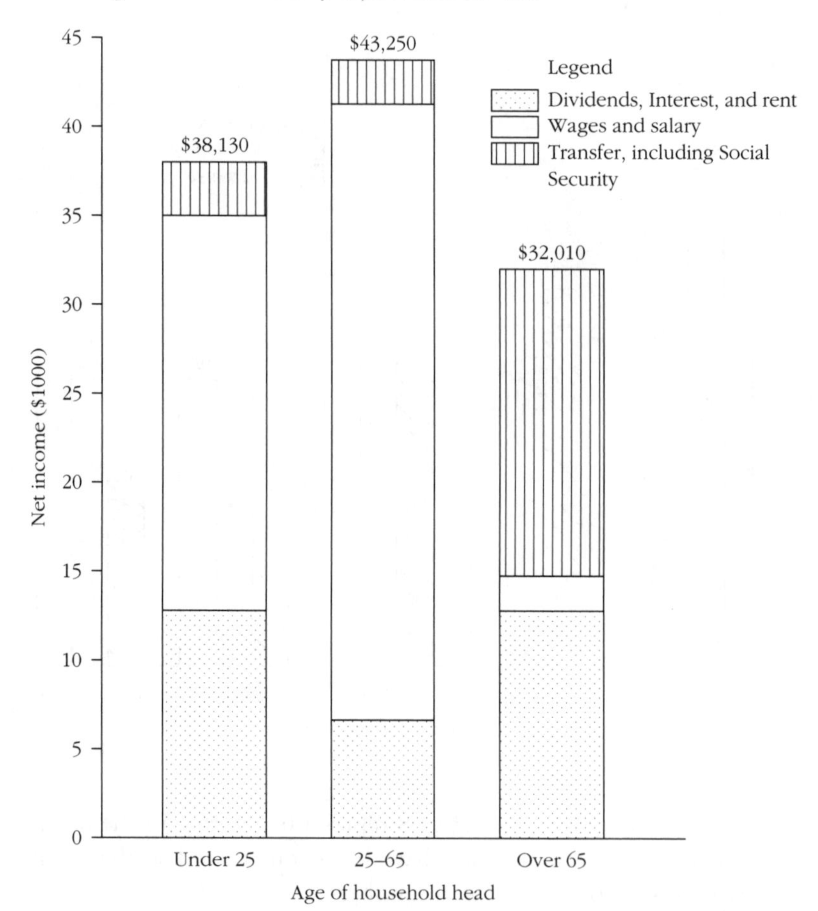

and graphics play a supportive role. The graphics can be relatively complex because people have time to study details.

In a verbal presentation, slides or transparencies can be the major focus but must be simpler than in a written report, with far less content. Dense graphics are extremely difficult to absorb. Often, people in the audience can't make out the details and, in any case, don't have time to understand the meaning.

The only place we recommend using simple tabular displays straight off the computer is in an appendix to the written report. These generally contain abbreviations, extra information, small type, and are simply not formatted for easy comprehension.

Consistency or Variation?

Several principles should guide the use of graphics. First, people can grasp more information if it is presented in a consistent format. In general, it is better to start with one type of graphic, and continue using it to illustrate a whole series of related questions. This is preferable to making people shift from rows to columns, bar to pie charts, or letters to numbers.

The second—seemingly contradictory—principle is that a varied format encourages people to remember an especially important point. When used sparingly, variation is a powerful way to get people's attention. Sometimes, we use a pie chart to illustrate the first, summarized results and then use a series of bar charts to provide detail. Only when we want the audience to switch their mental gears do we use a pie chart again. Other times we use one format up until the last, concluding point and then switch the basic display enough to make the information stand out and be remembered.

Slides or Overhead Transparencies?

If you are short on time or money, overhead transparencies are definitely the best option. They can be made quickly and at little expense. In addition, it's possible to make last-minute changes even by hand. Blank transparency sheets are readily available in office supply and book stores for nominal cost.

If you have money and are not strapped for time, the choice is more complicated. Slides allow the use of bolder color and design. Generally, they work better for larger audiences and more formal presentations. The room where the slides are shown must be relatively dark for people to see them and the presenter usually has to look at the screen rather than the audience.

Using transparencies means the room can be lighter and the presenter can look directly at his or her listeners. Hence, there is usually a much better opportunity for discussion and interaction. Since large groups often can't see them as well, we tend to use overheads only for smaller audiences and less formal presentations.

Regardless of whether you use slides or overheads, be sure there is an extra bulb available for your machine. Otherwise, you might have to abandon your well-planned visual presentation and explain everything in words—a harrowing last-minute change to avoid at all costs.

Using color in slides and transparencies can add life to a presentation, but be careful! Using a color like pale blue for the text on a transparency can make it unreadable. Colors that clash can literally make an audience uncomfortable.

Summary

People spend time and money to do surveys because they need new information. Presenting that information verbally and in writing involves different skills and meets the needs of different audiences. Still, the two kinds of reports complement each other. Both help ensure that all the hard-earned results get communicated to people who can use them.

For more detail on reporting results, see:

Chapter 18, "The Reporting of Survey Research," in *Survey Research Methods,* Second Edition, by Earl Babbie. Wadsworth, Belmont, CA, 1990.

The Visual Display of Quantitative Information, by Edward R. Tufte. Graphics Press, Cheshire, CT, 1983.

11

Advice, Resources, and Maintaining Perspective

Many years ago, one of us was in a situation similar to that of the city official mentioned in the preface to this book. A consultant had given the official a $50,000 bid to do a survey that would help decide whether to spend $25,000 on new street signs. In our case, we wanted to add a small deck to our home so we could put up a few lawn chairs. A well-intentioned professional designer explained how beautiful the backyard would be with a large, two-level deck instead of the small rectangle that we had in mind. Given our modest carpentry skills, the designer's idea was more complicated than we could handle. In any case, the cost of hiring someone to draw the plans and someone else to build the deck was more than we could afford.

Our solution was twofold. First, we reaffirmed that a small, simple deck would be adequate for our needs. Next, we bought a how-to book on carpentry with a good section on building decks. This book introduced us to the specific parts that go into a simple deck—ledgers, I-beams, spacers, and different kinds of nails and bolts. It also provided step-by-step instructions and explained essentials like how much space variously sized boards could span without support.

With all this information at hand, we carefully drew a diagram of the deck we planned to build. At the lumberyard, a helpful employee looked at our plans and explained our options in terms of materials. We bought what we needed and began building. Soon a carpenter friend stopped by and made a few suggestions about making the deck perfectly level. Today, our modest but good deck is still intact and useful, thanks to the book, a man at the lumberyard, and a little advice from a carpenter.

This real-life example illustrates what we wanted to accomplish in the preceding 10 chapters. Our goal was to lay out the essentials of what makes a good survey. In addition, we wanted to teach the skills needed to do relatively simple surveys and the ability to recognize when professional help is required. We also hoped to explain enough basic concepts and terminology so that, when faced with uncertainty, readers could ask for help from qualified professionals.

In this final chapter we help prepare readers to get professional help and stress the importance of inventorying resources. Finally, we make concluding comments about conducting surveys in a changing society.

Preparing Yourself to Ask for Advice

If we've been successful in writing this book, reading it will have prepared you to ask for help with your survey. Still, we suggest that you organize your thoughts in two important areas before actually picking up the phone to request advice from someone. First, recall what you know about the four sources of error (especially in the context of your own project), and second, think through the interrelated decisions that must be made in the course of doing a survey.

Revisit the Four Cornerstones

Sample surveys are unique in that they allow us to estimate population characteristics based on a relatively small fraction of the population. There are many other useful research methods, including focus groups, case studies, and informal interviews without a questionnaire. None, however, has this unique capability for allowing us to estimate population characteristics.

It should be clear by now that accurately estimating population characteristics such as what percentage of the households in your community own personal computers, requires:

- drawing a large enough sample of respondents, randomly, so that sampling error is kept to an acceptable level;
- selecting a sample of people in such a way that virtually all members of the population in which we are interested have an equal (or sometimes known) chance of being selected, thus, limiting coverage error;
- writing questions and arranging them in a questionnaire in ways that help avoid measurement error; and
- obtaining a response rate high enough to lessen concern about nonresponse error.

Seldom can one achieve perfection in any of these areas. Sample surveys always have a certain amount of sampling error (unless all members of the population of interest are asked to respond and actually do). Frequently, lists from which samples are drawn either include ineligible respondents or exclude people who should be included. Writing questions that elicit absolutely accurate answers is often impossible. And finally, virtually no survey achieves responses from every single person in the sample. Hence, we always talk about trying to make close estimates rather than being as exact as a finely calibrated thermometer or clock.

If we want to make *good* estimates based on a survey, we won't have the luxury of ignoring any one of the four cornerstones. Our best chance comes from focusing carefully and persistently on all four sources of error at every step of the survey.

As materials for building a deck, nails cannot substitute for bolts, nor boards for concrete footings. Each plays a critical role in making the final product solid and useful, and so do the four cornerstones of a good survey. We hope that in the future, you will seriously question the survey sponsor who says, "We interviewed 1,100 people, so our results are accurate within three percentage points" or "We got a 90 percent response rate so our findings are extremely reliable." A good survey has minimal error from all *four* sources: coverage, sampling, measurement, and nonresponse!

Think through the Big Picture

When people come to us for help, the first thing they often ask is whether a particular list of questions will work as an opening to their questionnaire. Unfortunately, we can't respond unless we see their cover letter and learn whether it connects reasonably to the first questions. Even then, we can't say much about the cover letter unless we know why the survey is being done in the first place. In short, we can be most helpful when we understand the whole framework, including all the elements of the survey in the context of the four error sources.

Hence, before asking for help, we suggest that you try answering the following questions:

- What is the purpose of the survey? What information is needed? How will it be used? (Chapter 3)
- What method seems to be the best suited for this particular survey? (Chapter 4)
- How should the sample be selected? What kind of coverage problems are likely to occur? (Chapter 5)
- How large should the sample be? (Chapter 5)
- What specific questions will be asked? (Chapter 6)
- How will the survey be implemented to get the highest possible response rate? (Chapter 8)
- In what form will the final results to be presented? (Chapter 10)

What advice you need will become clearer when you try to answer these questions. Also you'll be able to explain how the various parts of the survey fit together, making your advisor's job much easier!

Where to Go for Help

We live in a specialized world where knowing when and how to get help is often critical. Surveys are no exception. Hence, throughout this book, we have suggested where readers can get more specific information about various topics.

If you need an advisor, the nearest college or university is probably the best place to start looking. Unfortunately, there is no single, generally recognized and supported discipline of survey research so we can't recommend one department without qualification. Try contacting faculty members (and in some cases students) in departments of math, statistics, sociology, marketing, political science, or education. Both students and faculty with such backgrounds may be able to refer you to someone outside their department who might be helpful.

As you look for an advisor, keep in mind a very important point: Different disciplines emphasize different error sources, often at the expense of others. Statisticians are especially adept in the areas of sampling and coverage, but may know less about the causes of measurement and nonresponse error. Sociologists, psychologists, and others are more often trained to deal with measurement and nonresponse issues, but not with sampling and coverage.

Hence, it is important to know that you might get unbalanced advice. For example, someone might overemphasize sampling error, encouraging you to increase sample size to the extent that sampling error is reduced to tenths of a percent. At the same time, they might ignore other sources of error completely. Someone else might help you use dozens of questions to develop intricate scales on a particular topic, but then ignore how a very long questionnaire might affect response rates. Or they might suggest limiting your sample size to a very small number to keep data-collection costs down.

Recognize also that survey methodologists differ in their experience and comfort with each of the methods. Some are likely to have experience with mail surveys but not with telephone, or vice versa. Those whose skills are limited to one method may not be able to help when another method is required.

Organizations and Publications as a Source of Advice

Throughout this book, we have recommended many publications that provide more detailed and, in some cases, advanced material on how to conduct surveys. In addition, certain professional organizations regularly publish results of studies intended to improve survey methods. These associations and their journals include:

- The American Association for Public Opinion Research, and its journal, *Public Opinion Quarterly*.

- The American Statistical Association and its journals, *Journal of the American Statistical Association, The American Statistician,* and *Chance.*
- The American Marketing Association and its journal, *The Journal of Marketing Research.*

International journals published in English and specializing in survey methods are the *Journal of Official Statistics,* published by Statistics Sweden, and *Survey Methodology,* published by Statistics Canada.

Many other disciplinary and professional associations also publish journals that occasionally contain articles about survey research. These include sociology, political science, economics, education, psychology, nursing, community development, recreational planning, and business management.

Two extensive bibliographies on mail survey research are available. For work published before 1974, see the appendix to *Mail and Telephone Surveys: the Total Design Method* (Dillman 1978). For the years 1975 to 1989, see *Mail Surveys, A Comprehensive Bibliography* (Dillman and Sangster 1991).

The most extensive bibliography on telephone surveys currently available is included in the book *Telephone Survey Methodology* (Groves et al. 1988). In addition, many textbooks mentioned throughout this book contain useful, general bibliographies on survey research.

Take Stock of Your Resources and Plan Ahead

Even with good advice a survey does not happen all at once. Each of the ten steps to success outlined in Chapter 1 requires diligence and planning.

Some people think that doing a survey is something like the World Series in baseball, where it is only necessary to win four of seven games to be a champion. More appropriately, we think, a survey is like an automobile assembly line, where parts come together from multiple sources, and get connected to one another. Every part must be positioned and fastened correctly, or the final product won't work properly. At each step in a survey, from choosing objectives to presenting results, something critical has to happen. A failure at any point can be the fatal flaw that reduces your survey to irrelevancy.

Many people who have done surveys agree in principle with what we said above. Still, in explaining why their survey didn't work as well as expected, they give reasons like the following:

> I had no idea it would take 10 drafts of the questionnaire to get it approved by the advisory committee. That only left one

month to collect the data and write the report! That's why we didn't have time for any follow-up phone calls.

We ran out of money. We *couldn't* send out a replacement questionnaire!

One office sent out the questionnaires and another volunteered to mail the reminder postcards. The one doing the postcards got them done early, so they sent them out before the questionnaire!

I know we shouldn't have presented our results to the planning commission without visuals, but the computer work wasn't done until the night before. Well, at least we had a set of printouts for the staff.

In all of these situations, the survey organizers knew what had to be done but couldn't carry out their plans. For this reason, our final advice to our readers is to develop a work plan and timetable that budgets and coordinates all of your resources—including money, any volunteer efforts, and time for each activity that needs to be done.

When we encourage people to develop a budget and work plan, some respond with typed budgets and time schedules in a loose-leaf notebook. Others produce elaborate charts to post in their office. Still others prepare handwritten lists. The form isn't important but the content is. When they review their plans, these people often realize that unless they get more money they'll have to cut their sample size. Or alternatively, unless they revise their timetable, they will never get their results in time.

These people ultimately conducted successful surveys, because they made plans, realized potential problems, and made revisions before it was too late. The message is, plan ahead!

Practical Surveys in a Changing Society

When a friend, who is also a survey professional, learned we were writing this book, she asked, "Should people even consider doing their own survey? Isn't doing a survey so complex that it should be left to the professionals?" Our response was equally straightforward. "In the late twentieth century, surveys are too important to be left only to professional survey methodologists!"

When historians write of the great transitions that have occurred in this century, one of the major themes will certainly be our passage into the Information Age. In the past, we lived in a society in which communication was slow, access to information was limited, and our ability to process and use information was constrained by technology. It was a society in which long lead times were needed to publish information, in which attitudes changed slowly, and opportunities to solve problems had long planning horizons.

In those days, surveys were something special. They were done only on matters of great importance, and usually only for large groups and organizations. The planning horizons were long, the collection of data slow, and the analysis and publication of results proceeded at a snail's pace.

We now live in the Information Age. The marriage of two innovations—powerful, inexpensive computers and the capability to transmit huge amounts of information quickly—provide the basis for this important societal transition. Ideas and information of all types can be transmitted from nearly any place to anywhere in a matter of seconds. Attitudes change quickly, people expect quick responses to new problems, and long planning horizons are not tolerated. People can get information from multiple sources and can combine and reorganize it to solve practical problems.

Efforts to do surveys, now so common in our society, are both a product of and a reflection of the Information Age. Developments in computers and telecommunications make possible rapid survey design, handling of sample lists, data collection, and data processing, in time periods that would have been unthinkable less than a decade ago. These technologies are available to individuals and small organizations as well as large ones.

One can hardly stay in a hotel, attend a conference, subscribe to a magazine, or belong to an organization without being offered the opportunity to complete a survey. Increasingly, all types of organizations want to know what their customers want and how satisfied they are with services. Cities and counties want to know how services are received, and voluntary organizations want to know what their members need. In all these cases, demand is fueled by the uniqueness of the survey method—the ability to closely estimate population characteristics, usually by surveying only a small fraction of the entire population.

If you have reached these final pages and decided that doing a survey is not for you, we hope this book has helped you understand why some other course of action is desirable. On the other hand, if you have decided that a survey does meet your needs and that you have all the necessary resources, we congratulate you on undertaking what more and more people have found helpful in meeting their information needs. Best wishes for conducting your own survey!

References

Babbie, Earl. *The Practice of Social Research.* 4th ed. Belmont, CA: Wadsworth, 1986.

———. *Survey Research Methods.* 2d ed. Belmont, CA: Wadsworth, 1990.

Bailey, Kenneth D. *Methods of Social Research.* New York: The Free Press, 1987.

Beimer, Paul. *Measurement Errors in Surveys.* New York: John Wiley & Sons, 1991.

Bishop, George F., Hans-Juergen Hippler, Norbert Schwarz, and Fritz Strack. "A Comparison of Response Effects in Self-Administered and Telephone Surveys." In *Telephone Survey Methodology,* edited by Groves et al. New York: John Wiley & Sons, 1988.

Center for Community Change. *Searching for the Way That Works: An Analysis of FmHA Rural Development Policy and Implementation.* Washington, DC: The Aspen Institute, 1990.

Center for Rural Affairs. *Half a Glass of Water.* Center for Rural Affairs: Walthill, NE, March 1990.

Cochran, William G. *Sampling Techniques,* 3d ed. New York: John Wiley & Sons, 1977.

Dillman, Don A. *Mail and Telephone Surveys: The Total Design Method.* New York: John Wiley & Sons, 1978.

———. Chapter 11: "Elements of Success." In *Needs Assessment: Theory and Methods,* edited by Donald Johnson et al. Ames, IA: Iowa State University Press, 1987.

———. "Our Changing Sample Survey Technologies." *Choices* 4 (3): 12–15 (1989a).

———. "Response to Fesco." *Choices* 4 (4): 40–41 (1989b).

———. "The Design and Administration of Mail Surveys." *Annual Review of Sociology* 17: 225–249 (1991).

Dillman, Don A., Jean Gorton Gallegos, and James H. Frey. "Decreasing Refusal Rates for Telephone Interviews." *Public Opinion Quarterly* 50 (1): 66–78 (1976).

Dillman, Don A., and Roberta L. Sangster. "Mail Surveys: A Comprehensive Bibliography: 1974–1989." *CPL Bibliography 272.* Monticello, IL: Council of Planning Librarians, 1991.

Dillman, Don A., and John Tarnai. "Mode Effects of Cognitively-Designed Recall Questions: A Comparison of Answers to Telephone and Mail Surveys." In *Measurement Errors in Surveys,* edited by Paul Biemer et al. New York: John Wiley & Sons, 1991.

Fowler, Floyd J., Jr. *Survey Research Methods.* 2d ed. Newbury Park, CA: Sage Publications, 1993.

Frey, James H. *Survey Research by Telephone.* 2d ed. Newbury Park, CA: Sage Publications, 1989.

Gleick, James. "The Census, Why We Can't Count." *New York Times Magazine,* July 15, 1990.

Groves, Robert M. *Survey Errors and Survey Costs.* New York: John Wiley & Sons, 1989.

Groves, Robert M., and James M. Lepkowski. "Dual Frame, Mixed Mode Survey Designs." *Journal of Official Statistics* 1 (3): 263–286 (1985).

Groves, Robert M., Paul B. Biemer, Lars E. Lyberg, James T. Massey, William L. Nicholls, II, and Jospeh Waksberg, eds. *Telephone Survey Methodology.* New York: John Wiley & Sons, 1988.

Guenzel, Pamela J., Tracy R. Berckmans, and Charles F. Cannell. *General Interviewing Techniques: A Self-Instructional Workbook for Telephone and Personal Interview Training.* Institute for Social Research, University of Michigan, Ann Arbor: Survey Research Center, 1983.

Henry, Gary. *Practical Sampling.* Applied Social Research Methods Series, no. 21. Newbury Park, CA: Sage Publications, 1990.

James, Jeannine M., and Richard Bollstein. "The Effect of Monetary Incentives and Follow-Up Mailings on the Response Rate and Response Quality in Mail Surveys." *Public Opinion Quarterly* 54: 346–361 (1990).

———. "Large Monetary Incentives and Their Effect on Mail Survey Response Rates." *Public Opinion Quarterly* 56: 442–453 (1992).

Kachigan, Sam K. *Statistical Analysis: An Interdisciplinary Introduction to Univariate and Multivariate Methods.* New York: Radius Press, 1986.

Kish, Leslie. *Survey Sampling.* New York: John Wiley & Sons, 1965.

Krysan, Maria, Howard Schuman, Lesli Jo Scott, and Paul Beatty. Unpublished paper, "Mail vs. Face-to-Face Surveys: A Comparison of Response Rates and Response Content Based on a Probability Sample," University of Michigan, Ann Arbor, MI, 1993.

Kulka, Richard A., Nicholas A. Holt, Woody Carter, and Kathryn L. Dowd. "Self-Reports of Time Pressures, Concerns for Privacy and Participation in the 1990 Mail Census." Presented at the 1991 Annual Research Conference, Bureau of the Census, Arlington, Virginia, March 17–20, 1991.

Lavrakas, Paul J. *Telephone Survey Methods: Sampling, Selection, and Supervision,* 2d ed. Newbury Park, CA: Sage Publications, 1993.

Lepkowski, James M. "Telephone Sampling Methods in the United States." In *Telephone Survey Methodology,* edited by Robert M. Groves et al. New York: John Wiley & Sons, 1988.

Morgan, David L. *Focus Groups as Qualitative Research,* Vol. 16, *Qualitative Research Methods.* Newbury Park, CA: Sage Publications, 1988.

Nederhof, Anton J. "Effects of a Final Telephone Reminder and Questionnaire Cover Design in Mail Surveys." *Social Science Research* 17 (4): 353–61 (1983).

New York Times, August 26, 1990.

Polling Report, Inc. Vol. 8, No. 21, Washington, DC 20002, 1992.

Salant, Priscilla. *A Community Researcher's Guide to Rural Data.* Washington, DC: Island Press, 1990.

Schuman, Howard, and Stanley Presser. *Questions and Answers in Attitude Surveys: Experiments on Question Form, Wording, and Context.* New York: Academic Press, 1981.

Shettle, Carolyn F. "Evaluation of Using Incentives to Increase Response Rates in a Government Survey." Presented at 1993 Joint Statistical Meetings, American Statistical Association, San Francisco, August 1993.

Sudman, Seymour, and Norman M. Bradburn. *Asking Questions: A Practical Guide to Questionnaire Design.* San Francisco, CA: Jossey-Bass Publishers, 1982.

Survey Research Center. *General Interviewing Techniques: A Self-Instructional Workbook for Telephone and Personal Interviewer Training.* University of Michigan, Ann Arbor: Survey Research Center, 1983.

Thornberry, Owen T., and James T. Massey. "Trends in United States Telephone Coverage across Time and Subgroup." In *Telephone Survey Methodology,* Groves et al., eds. New York: John Wiley & Sons, 1988.

Voss, Paul, Stephen Tordella, and David Brown. "Role of Secondary Data." In *Needs Assessment, Theory and Methods,* edited by Donald Johnson et al. Ames, IA: Iowa State University Press, 1987.

Weber, Robert Phillip. *Basic Content Analysis.* Sage University Papers Series, No. 07-049. Beverly Hills, CA: Sage Publications, 1985.

Yin, Robert K. *Case Study Research: Design and Methods.* Newbury Park, CA: Sage Publications, 1989.

Index

A

B

C

D

E

F

G

H

I

L

M

N

O

P

Q

R

S

T

U

V

W